Dr Vinod Kotwal belongs to the 1994 batch of the Indian Civil Services and is currently posted as Member Secretary, National Pharmaceutical Pricing Authority. She has written two poetry books—*Twilight Wanderings* and *From Within* and a book on Codex Alimentarius Commission, a joint FAO-WHO body. Vinod has also written multiple articles, papers and monographs for prestigious journals and magazines on wide ranging subjects from food safety to ethics in artificial intelligence.

Looking for Jeetu

The Story of a Braveheart—
Major Shatrujeet Kotwal, KC

VINOD KOTWAL

SPEAKING TIGER

SPEAKING TIGER BOOKS LLP
125A, Ground Floor, Shahpur Jat, near Asiad Village,
New Delhi 110049

First published by Speaking Tiger Books in 2024

ISBN: 978-93-5447-863-5

10 9 8 7 6 5 4 3 2 1

Typeset in Adobe Jenson Pro by SÜRYA, New Delhi
Printed at HT Media Ltd, Greater Noida

Pankaj Sharma, Hemant Bisht, Chaitanya Marathe—
Jeetu's NDA batchmates who also left us too early

Late Major K.P. Vinay, KC, Major Raghu Raman, AC
and his colleagues in the 34 RR

Officers and soldiers of the 3 Jat Regiment

The Band of Brothers who stuck together through thick
and thin and remember Jeetu fondly

Contents

Prologue

Heart, we will forget him!
You and I, tonight!
You may forget the warmth he gave,
I will forget the light.
When you have done, pray tell me
That I my thoughts may dim;
Haste! Lest while you're lagging
I remember him!
—Emily Dickinson
(Composed 1858, published 1896)

I began to pen down these words on 15 January 2023, though this book has been in my head for the past fourteen years. My laptop has a one-page Word document, dated 23 April 2009, titled 'A Sister Remembers'. And that is all the text it contains—the rest of the page is blank. It reminds me of this line: 'In a sense, I have been writing this book for fifteen years,' from the last chapter of Rabbi Harold S. Kushner's book *When Bad Things Happen to Good People*.This slim book of around 150 pages was published in 1981, after the death of his fourteen-year-old son in the year 1977, the year my brother was born.

This book is about my brother, Major Shatrujeet Kotwal—Jeetu to the family—who died on 17 June 2008 in a road accident in Pune, a month and half short of his thirty-first birthday, achieving much in his short life and scaling many summits. In January 2008, he had been awarded the Kirti Chakra (KC), the second-highest peacetime gallantry award; he was working as an instructor at his alma mater, the National Defence Academy (NDA), Khadakwasla, Pune; and he was to leave for a United Nations mission in Congo in July 2008.

Charlotte Bronte writes in her poem 'Life':

> What though Death at times steps in,
> And calls our Best away?
> What though sorrow seems to win,
> O'er hope, a heavy sway?

Jeetu came into our life like a meteor and left as quickly, leaving behind a stream of stardust. Although my science training tells me that meteors are nothing but masses of dust and debris from space that burn in the earth's atmosphere, how do I explain the coincidence to my rational mind that Jeetu was born on 9 August and the Perseid meteor shower peaks around 9–14 August every year, leaving behind long swathes of light and colour as it blazes across the Earth's atmosphere?

Like a squirrel, I have been collecting bits of paper,

photographs and storing fragments of conversation in my head. I was unable to muster the courage to write his story for multiple reasons, the foremost being my own inadequacies as a writer—the fear that rather than celebrating his life, the story I would write would turn into a maudlin account of loss, sorrow and grief—together with this unceasing nagging thought, *what is the need?*

Two events or catalysts, whatever one may call them, finally made me embark on writing this book.

The first was a meeting in December 2022 with Colonel Dheeraj Kotwal, Jeetu's batchmate in the NDA as well as in the Indian Military Academy (IMA), Dehradun, and a very close friend. During our conversation, which centred mostly around Jeetu, he mentioned that their batch at the NDA was planning a twenty-five-year reunion in June 2023. He reminisced about their other batchmates too who were no more, particularly Captain Pankaj Sharma, also from Jammu. Colonel Dheeraj regaled us with stories of their camaraderie and brotherhood. The three of them—Jeetu, Dheeraj and Pankaj—were thick as thieves, travelling together back and forth in the NDA Express, the special train that runs every six months to ferry NDA cadets going north during their term-breaks from Pune. There would be a break in the journey at Delhi before they caught another train to Jammu in the evening. Dheeraj is a raconteur and vividly remembers visiting our Vasant Kunj home and the food they ate before the three caught their train in the evening to Jammu. Good food is the only thing that can tempt an NDA cadet—this I would gather during interactions with other batchmates of Jeetu's too. Dheeraj fondly remembered Pankaj's body-building endeavours, his childlike innocence and his very outgoing nature. They lost Pankaj in 2000 in a Line of Control (LoC) incident; he was barely twenty-three years old.

Pankaj was into bodybuilding while Dheeraj and Jeetu were into cross-country and long-distance running. They had their own strategies for being at the forefront in these runs at the NDA as well as the IMA. The strategies were explained in detail, as were Dheeraj's attempts to guard Jeetu's teacup lest he should knock it off while struggling to keep awake studying for exams in their Young Officers (YOs) course. They were roommates for almost five months while doing their YOs course in 2000 in Mhow.

Still, this was all oral history. I had read somewhere that oral history, in a sense, 'is a struggle of memory against forgetting'.

This fear clutched my heart. Will Jeetu's memories fade and become sepia-tinted photographs, blurring at the edges? Does his son, Kartikeya, who was born after his father's death, not have the right to know about his father?

Randy Pausch, professor of Computer Science, human-computer interaction and design at Carnegie Mellon University (CMU), Pittsburgh, died in July 2008, the same year as Jeetu, a few months short of his forty-eighth birthday. He was diagnosed with pancreatic cancer in 2006 and given a terminal diagnosis a year later. Knowing that he had only a few months left, he delivered his 'Last Lecture', titled 'Really Achieving Your Childhood Dreams', at CMU on 18 September 2007. Watching that video gives me goosebumps. This was also published as a book that Pausch co-authored with Jeffrey Zaslow. In the book, Pausch writes, 'I was trying to put myself in a bottle that would one day wash up on the beach for my children. If I were a painter, I would have painted for them. But I am a lecturer, so I lectured.' Randy Pausch continues: Who was my dad? What was he like? Hopefully, his book could give them the answers.

Our brother did not get the time to leave a message in a bottle or deliver a lecture for his son. His sudden and untimely death robbed him of that opportunity. He did not get the time to say his goodbyes. However, in his short but glorious life, he left behind a legacy that will serve as a beacon not only for his son but for all youngsters who believe in the power of hard work, striving for excellence, and the 'never say die' spirit.

The second event, the other catalyst if you may, was reading Pico Iyer's book *Autumn Light: Season of Fire and Farewells*. It's an exquisitely written elegy in prose on many things—the death of his father-in-law, old age, loss, pain and death—with autumn in Japan as a metaphor for the transience of life. I also wanted to put down on paper with urgency my brother's life that intersected and touched so many, although maybe not as beautifully as Pico Iyer.

With these thoughts swirling in my mind, I embarked

on writing this memoir, piecing together thoughts and memories to create something of beauty and strength. It reminded me of the Japanese art of *kintsugi,* also known as 'golden joinery' or 'golden repair'. It is the art of repairing broken pottery by mending the areas of breakage with lacquer dusted or mixed with powdered gold, silver or platinum. It is said that as a philosophy, *kintsugi* is similar to the Japanese philosophy of *wabi-sabi,* a worldview centred on the acceptance of transience and imperfection. 'Kintsugi encourages us to embrace our past along with its scars; to realize that our cracks make us even more beautiful', writes Pallavi Aiyar in her book *Orienting: An Indian in Japan,* a part travelogue, part social commentary on the country.

Patrick Lane, the Canadian poet, writes in his poem 'Kintsugi':

> There is only a little left to know. Water returns to water.
> The dew in my eyes, a moment ago an ocean.
> On the face of a golden puddle the moon repairs itself
> again and yet, again.

Maybe, it was time for us to accept our loss, heal ourselves by celebrating his life, and embrace our brokenness.

1

Roots of the Mountain Warrior

The hills will never stop calling us.
You just have to stop and listen.
—Bulbul Sharma, *Between Heaven & Earth: Writings on the Indian Hills*

Our parents, Shri Ram and Kiran Kotwal, trace their roots to Bhadarwah, a beautiful valley in Jammu and Kashmir, situated at an elevation of around 1,613 metres. It is also referred to as Chota Kashmir or Little Kashmir by many. Mountains are thus in the helical structure of our DNA and for me they are the siren songs I cannot ignore, maybe because I was born there. But Jeetu too had a special love and affinity for the mountains though he was born several thousand miles away from the land of his ancestors on 9 August 1977 at the Military Base Hospital, Narangi, Assam, where our father was stationed at that time. Jeetu was named 'Shatrujeet' by our mother, which is Sanskrit for the one who vanquishes the enemy. I guess he lived up to his name.

It is believed that Bhadarwah is made up of two words, 'bhadar'and 'wah'—meaning a place where people are gracious and kind. The river Neru and other smaller gurgling streams course their way through the saucer-shaped Bhadarwah valley, which revels in different colours and hues as seasons change. During winter, it is carpeted with snow and then in spring it is ablaze with dahlias, roses and lilies. Then there are the pink and white blossoms of apple, pear, almond, apricot, plum, and cherry—delicate trees draped in candy floss colours—contrasting with the sturdy walnut trees with their long compound leaves. High in the mountains are the majestic deodar and kail (Himalayan cedar and Himalayan pine, respectively, or *Cedrusdeodara* and *Pinus wallichiana*). With lush green paddy fields and the snow-capped mountains in the background, Bhadarwah can beat any alpine village in beauty.

According to some, Bhadarwah traces its existence to the Mahabharata days. To its east, on the bank of the Neru, there is the Gupt Ganga temple dedicated to Lord Shiva. Inside the temple you will find a stone with a footprint believed to be that of Bhima, the second-oldest of the five Pandava brothers. An adjoining tunnel close to the temple is believed to have helped the Pandavas escape to an unknown destination.

Bhadarwah is a region steeped in traditions, folklore,

fairs or melas, and festivals, including the Mela Pat and the Sobar Dhar Mela, where the echo of gentle lilting Pahari songs full of love, pathos and yearning, waft through the valley. There are no robust, vigorous dances like the bhangra or gidda of Punjab, just gentle swaying to the mellow music in a circle. Even non-dancers like me can sway to the melodic tunes without being self-conscious.

Apart from its rich cultural traditions, Bhadarwah also has high literacy rates that can match those of Kerala. It was always a matter of wonder for me how a small and remote valley that, till the 1980s, would get cut-off from the rest of the world during winter, was able to achieve this feat. In the early 1960s, our father was sent to Jammu to study for his bachelor's degree in science and our mother to Srinagar for a two-year diploma course in health and family welfare. I have not come across any instance of gender discrimination as far as imparting education is concerned. Equal opportunities were given to my uncles and aunts. Most of my close relatives and cousins are postgraduates. Education is paramount and has been a priority not only for my family but for the region as a whole.

I remember my paternal grandmother, Daadi, coming to visit us from Bhadarwah during the winters in the late 1980s with an armload of daffodils, guchhi mushrooms, and an edible fern known as kasrod among other goodies from the mountains, including dried beans, sundried tomatoes, and banafsha flowers (Himalayan wild blue violet, or *Viola odorata*). She was a great cook, rustling up batches of pumpkin flower fritters; kashli or steamed pumpkin wedges, sprinkled with sugar if the pumpkin was not sweet enough; goat trotters pulao...Daadi used to make a flavourful pulao from guchhi mushrooms too. I still recall the musty earthy flavour of the guchhis and the stories Daadi would recount. These mushrooms are not cultivated commercially; instead women in remote mountainous villages would collect them from the coniferous forests and bring them down to the Bhadarwah valley and then barter them for clothes. Much later in life, we realized that guchhi mushrooms are expensive and highly sought after not only for their earthy flavour but also as an ingredient in gourmet cuisine.

Daadi's cooking was effortless and without fuss. Cooking for a big family of eight children, a finicky husband, and a constant stream of relatives had given her enough practice. Kasrod pickle made in mustard oil would make our noses water, much like wasabi does. Imagine my delight when decades later in 2016, I came across kasrod clinging to a hillside in the town of Sorrento in Italy's Amalfi coast. My family was amused at my reaction and the ferocity with which I was clicking pictures of kasrod from various angles to share with my mother, rather than looking at the famed Sorrento lemons.

Daadi always wore long printed kurtas and plain soothans (churidar pyjamas with more rings than the ones we wear) with a plain white mulmul dupatta draped over her head. Though she was named Saraswati, the goddess of learning, Daadi stopped going to school after class five, mainly because of the caning she would receive from her schoolteacher. Yet, she would always tell me and my sister to study hard and get a job. 'Never depend on your husband for money. Your Daadaji was a good man but I never liked it when I had to ask him for a few annas to buy dye to colour my dupatta,' she would say.

Daadaji was a revenue officer in the state government but this proud woman would grow vegetables in her kitchen garden—long hill cucumbers, brinjals and tomatoes—and then send her servants to sell the surplus in the local market. What Daadi lacked in education, she made up for by her wisdom of having lived a life that was tough and tumultuous, but fulfilling. Unlike the popular perception of strained relations between mothers-in-law and daughters-in-law, mostly fuelled by the media and TV soaps, Daadi and Mom got along like a house on fire. Two beautiful, strong, opinionated women with their own strengths and weaknesses. Jeetu was brought up in a household where such strong women were at the forefront.

Body as well as soul were nourished by Daadi's gifts. The multi-layered imagery in Ranjit Hoskote's poem, 'Portrait of a Lady'—written with his grandmother in mind, somehow captures the essence of my grandmother too:

> Objects are lessons; from bowls, hairpin, brooches,
> you yearn of forgotten lives. The stories say

my grandmother was a fever tree:
Two birds sat on her branches, one pecking
At a grape, the other singing an aria.

Our ancestral home in Bhadarwah is perched on a hillock, below which flows a stream that we call 'gad'. During spring, the slopes of the hillock are covered with a blanket of daffodils, quite reminiscent of William Wordsworth's eponymous poem. Daffodils to me are synonymous with my Daadi, a beautiful and strong woman. Getting daffodils is rare in Delhi and during winter, whenever they are available with the neighbourhood florist, I buy all the bunches hurriedly. The florist is quite confused at my eagerness! Reverently, I put them in a glass vase and then place the vase on my study table. I never tire of looking at them. And when they wilt, I cannot bear to throw them in the dustbin; instead I scatter them among flower pots.

The imagery of daffodils in William Wordsworth's poem is more widely known. But Robert Herrick's poem 'To Daffodils' is no less poignant:

Fair Daffodils, we weep to see
You haste away so soon;
As yet the early-rising sun
Has not attain'd his noon.
Stay, stay...
As you, or anything.
We die
As your hours do, and dry

Most of Herrick's poems were inspired by the Latin expression '*carpe diem*' (seize the day) and focus on the transience of life as well as enjoying each present moment, akin to 'mindfulness' as we understand today. This was expressed way back by this seventeenth-century English poet, and many years later, while talking to his batchmates, I would find that Jeetu also followed the same philosophy—living in the present.

The Neru's source is Kailash Kund, situated at around 4,265 metres. It is also the source of the River Tawi. Kailash Kund is believed to be the original abode of Lord Shiva, which he left to Vasuki Nag before moving to Manimahesh in the Bharmour area of Himachal Pradesh. Bhadarwah is also called 'nagonkibhoomi', the land of snakes, because it is home to many varieties of snakes. Our kuldevi, family deity, too is Nagini Devi, the snake goddess. A small, ancient temple, up in the village of Malani, in a meadow surrounded by tall fir trees, is dedicated to her.

The Kailash yatra, the annual three-day pilgrimage dedicated to Vasuki Nag, the deity of Bhadarwah town, is held in August or September and commences on the fourteenth day after Shravan Purnima. The yatra begins at the ancient Vasuki Nag temple in Gatha village with the charrimubarak or the holy mace being taken out from the temple. It is joined by another mace from the Vasuki Nag temple at Vasik Dhera in Bhadarwah town.The twenty-one-kilometre yatra then commences, passing through a series of quaint hill villages, the lush Seojdhaar meadow and coniferous forests, finally descending to the rugged, boulder-filled plains at the base of Mount Kablas (Kailash). The idol of Nandi, the sacred bull of Lord Shiva, at the mouth of the dark grey waters of the Kailash Kund where the temple is built, is a silent witness to devotees paying their obeisance to the Almighty for centuries.

I have not undertaken the Kailash yatra yet, but I remember Jeetu doing so in 2005 with Ranjeet, his close friend since school days, and Venkat, our paternal aunt's son. There are a few pictures of all three of them there. Mom and Abha, Jeetu's wife, had stayed back at Bhadarwah. Mom still goes into raptures reminiscing about growing up in Bhadarwah and her participation in the Kailash yatra in the late 1960s or '70s. According to her, the grass in Seojdhaar is as sweet as sugar and the air intoxicating with the fragrance of herbs, flowers and trees. Mom's description of the Seojdhaar used to transport me to Johanna Spyri's alpine meadows in her book *Heidi*.

Barely three or four months after his birth, Jeetu had fallen seriously ill with pneumonia. He was admitted to the same hospital where he was born. Seeing her son hanging on to life in the twilight zone between life and death, Mom, in desperation, invoked all the gods and prayed to Lord Shiva: 'Oh Lord, save my son and I will bring him to your abode to pay obeisance.'

Jeetu recovered, and the promise made to the gods had to be kept. Mom took her young son, who was just seven or eight years old at that time, on the arduous Manimahesh yatra in the month of Bhadon in 1984. The Manimahesh Lake, 26 km from Bharmour in the Budhil valley of the Chamba district in Himachal Pradesh, is situated at about 3,960 metres, at the foot of the 5,658-metre-tall Kailash peak. A rock formation in the shape of a shivling at the peak is considered to be a manifestation of Lord Shiva. At one corner of the lake is a marble statue of Lord Shiva, which is worshipped by the devotees after taking a dip in the holy lake.

Mom has always been a great storyteller and she has an innate ability of bringing alive events, conversations and encounters with the right amount of drama. Whenever Shravani, my daughter, wants to hear family gossip, she prefers to go to her grandmother. Shravani never ceases to tell me, 'Mom, you are so boring; Naani is so good at narrating things.' Her Naani reminds her of the granny in the serial *Downton Abbey*, the ever curious and indomitable Lady Violet Crawley.

After nearly forty years, I still vividly remember Mom bringing alive the Manimahesh yatra for us, and telling us how Jeetu had trekked all the way without getting on to a mule or a porter. He did not complain even once during this arduous journey. One of my paternal uncles, Ramprasad Chachu, who had accompanied them, was awestruck by little Jeetu's stamina and endurance. There was a streak of spirituality tempered with perseverance in Jeetu. On one of Jeetu's visits to our home in Delhi, we had had an intense discussion on religion, philosophy and the faith that drives us. He had then told me about his interest in Kundalini Yoga. I was a little surprised. Here was this fun-loving young man in his twenties, a movie buff, thinking so deeply on such issues.

Falling prey to anthropogenic factors, Bhadarwah is also witnessing unchecked growth. Its paddy fields are now dotted with concrete structures and its streams are choked with plastic waste. For how long it will remain 'untouched' and true to the tagline—'Heavens untouched'—of the Bhadarwah Development Authority, created in 2005, remains to be seen. The fragile ecosystem of the young mountains is in peril, and maintaining the right balance between development and sustainability may be a tad difficult, though not insurmountable. The unfortunate developments in Joshimath is a wake-up call for all of us, and recently it was reported in the media that a few houses in a village in the Doda region too had developed cracks.

Jeetu was very sensitive to environmental issues and keen on protecting nature. Lieutenant Colonel Rohan, an artillery officer and Jeetu's younger colleague in the 34 Rashtriya Rifles (RR), from 2005-2007 when they were stationed in Budgam in the Kashmir valley, vividly remembers that on a few occasions while patrolling at night, deep in the dense coniferous forests, they had witnessed illegal tree felling and smuggling of timber. They had seen about a hundred or more mules standing ready to ferry the illegally felled deodar and fir trees—a sight that would make Jeetu fume: 'How can they destroy jungles like this? Cutting trees without permission is illegal. We are custodians of these forests, and should protect them and pass them on to the next generation. Such offenders should be punished.' Heeding to their commanding officer's advice, the matter was duly reported to the civilian authorities, who usually handled such issues. Jeetu was not the one to turn a blind eye to wrongdoings.

I guess with our ancestral roots in the mountains, there is a natural affinity towards nature; we have always internalized the stillness and the sturdiness of the mountains. This I think helped Jeetu to do exceptionally well in the extremely tough and arduous Mountain and Winter Warfare courses that he completed at the High Altitude Warfare School (HAWS).

A few days ago, while surfing the internet, Madhu, my husband, came across a documentary made by the Discovery Channel on HAWS, the most famous among the warfare academies of the world. The school is in Gulmarg, Jammu and Kashmir, and was established as a defence service training and research establishment in 1948 by General K.S. Thimayya. It is globally acclaimed for training Indian soldiers for operating at high and super-high altitudes. Soldiers from other countries also visit the school for specialist training.

Watching that documentary made me realize the

limits to which you had to push yourself physically, psychologically and emotionally to be a part of the school. The fact that Jeetu did consistently well in all the four courses was a revelation to me. How my stoic brother never ever complained about the hardships of the courses! Instead, he only spoke about the experiences. I guess this is how they train you at the NDA and the IMA—the triumph of mind over matter.

2

The Growing-up Years

Though we travel the world over,
To find the beautiful,
We must carry it with us,
Or we find it not.
—Ralph Waldo Emerson

Udhampur is the headquarters of the Northern Command of the Indian Army, which was originally formed as the Northern Army of the British Indian Army in 1908, although that branch was scrapped in 1947 and the unit re-raised in 1972. Currently, the XIV Corps (Leh), XV Corps (Srinagar), I Corps (Mathura) and XVI Corps (Nagrota) are under the control of the Northern Command. The residence of the General Officer Commanding (GoC) of the Northern Command is on Tara Mount, a little hillock named after Tara Devi, the fourth wife of Shri Hari Singh, the erstwhile ruler of Jammu and Kashmir. The hillock overlooks the army cantonment spread across.

Papa took voluntary retirement from the army and decided to settle down in Udhampur. A number of factors favoured this decision. Mom was working in the Health and Family Welfare Department, Jammu and Kashmir government, and was posted here, and our parents wanted their three children (I am the eldest and Jeetu, the youngest, with Vinkoo, our sister, between the two of us) to get a good education. Udhampur had good schools, including a Kendriya Vidyalaya and the Army Public School, where Jeetu studied and completed his schooling in 1995. Vinkoo and I were students of the Kendriya Vidyalaya.

Udhampur also had an army club with a swimming pool and a squash court. Vinkoo recalls: 'We started swimming quite young. During summer breaks, Jeetu and I used to go every day to the swimming pool and swim for a good couple of hours throughout the summer break. He was a quick learner and mastered all the strokes swiftly. We used to be famished after swimming and devoured bourbon biscuits, fruit and at times made 'panjiri'. From his childhood, Jeetu was very lean, had an athletic build and never put on any weight. He loved cricket, cycling, swimming and running.' Ranjeet echoes similar thoughts: 'Shatrujeet was a very good sportsman. He excelled in the track and field events in school and was also a long-distance runner. We used to play inter-house and inter-section matches. We were opponents in games as he was in Beas House and I was in Ravi House.'

Another attraction of Udhampur was the wonderfully well-stocked Northern HQ library. It was our 'magic land'. As Emily Dickinson, one of my favourite poets puts it, 'There is no frigate like a book / To take us Lands away.' For the three of us, books were the frigates and the planes to carry us across continents and oceans. There was also the Chinar Cinema in the Chinar compound, our portal to another world of visual fantasy, traversing the realms of English as well as Hindi movies.

So, rather than settling down in Udhampur city, our parents bought a piece of land in Kotli Pain village, from which all these facilities were within walking distance. The house was built between the town and the village, with the army camp on one side and wheat and mustard fields on the other. There was a gnarled mulberry tree at one corner of the plot that shared its boundary with a field of wheat and mustard. I recall Jeetu climbing the tree; many years later, Shravani would also climb the tree, scraping her knees. A small pine sapling was planted in the front lawn and it now towers over the house, its branches rustling furiously in the famous, or infamous, Udhampur winds, shedding pine needles in copious amounts. Around the middle of 1988, Arjun Sharma, a young boy from a poor family, came into our life and continues to be a part of it. He lived with us and helped around the house; our parents got him admitted to the school. He looked up to Jeetu as his elder brother.

We watched classics like *Ben-Hur, Chariots of Fire,*

Gone with the Wind, *The Bridge on the River Kwai*, *Papillon*, and *Kramer vs. Kramer* in the Chinar Cinema on Wednesdays, while gorging on greasy packs of popcorn and soft drinks. While walking back home, our mood used to depend on the movie that we had watched—sometimes pensive after watching the Hindi movie *Dosti*, at other times excited, discussing the chariot race in *Ben-Hur* or going gaga over Rhett Butler. This was our exposure to world cinema. Some puritans may say it was pure Hollywood kitsch and not representative of world cinema at all. We did not care, for these were magical moments for us, conjuring wondrous places, events and people. In hindsight, I think Jeetu's abiding love for movies stemmed from this. He would not lose a chance to watch a film. In 2006 or 2007, during one of their visits to Delhi, four of us—Abha, Jeetu, Shravani and I—had gone to watch the 11.45 p.m. show of the James Bond movie, *Casino Royale*. We came back home at around 3 a.m. That is the only late night show I have seen till date, and I could only do so with my 'bindaas' brother.

Jeetu was particularly attached to Udhampur and there were constant arguments between him and Mom on the relative merits of a bigger lawn and a kitchen garden. Jeetu had plans to construct an additional floor with a glass-enclosed library atop the single-storey house our parents had built.

Dheeraj has recently built a house just behind ours in Kotli Pain and he tells me, 'This is my tribute to my friend and cousin—army people know us as cousins. I would want to spend my old age where he spent his childhood.'

While growing up, we were taught to cherish knowledge. We may not have had a wardrobe full of clothes but our home was always brimming with books and magazines. I distinctly remember Papa subscribing to the *Competition Success Review* and the *Reader's Digest,* when I was in the sixth standard. Their annual subscriptions were religiously renewed till the time Jeetu, the last in the family, cleared his NDA exam. In addition, there were the daily newspapers to read, one local and the other national. It would be evening by the time the national newspaper reached us, travelling by train till Jammu and then by road to Udhampur. The Uttar Sampark Kranti Express was the first train to be flagged off from Udhampur's quaint railway station in April 2005, twenty-three years after its foundation stone was laid by Indira Gandhi in 1982, the then Prime Minister. Till then, the only way to reach Udhampur was via a long journey from Jammu by road. So, unlike city dwellers who read their newspaper with a morning cup of tea, we read our national newspaper with our evening tea. Now, Mom has two newspaper apps on her smartphone and is more clued in on politics than both me and my sister. With all this reading material floating around us, the three of us never thought of studying separately for the general-knowledge paper while preparing for various competitive examinations.

The entrance examination for the NDA, considered to

be one of the toughest competitive exams in the country, is held twice a year. For about 360 vacancies, there are 2.5 lakh aspirants. After a written examination, 8,000-10,000 are shortlisted for the Services Selection Board (SSB), which is incredibly challenging because of its low success rate. Barely half a dozen candidates out of every hundred make it past the SSB interview, not to mention the SSB screening tests, in which sixty per cent of the candidates fail. 'Shatrujeet's overall preparation for the NDA entrance exam was not that strong; however, his general awareness and basics were far better than many. He appeared for the entrance exam twice and cleared both times,' recalls Ranjeet. In one of my recent trips to Dehradun and beyond, I was struck with the number of hoardings along the highway of institutes claiming to train young men for the NDA/CDS entrance examinations. And then there was my young brother, effortlessly clearing his entrance examinations without any coaching/training!

Jeetu wrote me a five-page-long letter—his longest one to me—on 25 February 1995, a day after his return from Allahabad after clearing his SSB. A total of seventy-two boys were divided into groups, and Jeetu was assigned to Selection Board 19. According to him, 'At Allahabad selection centre east, there are four SSBs, namely 11, 19, 18 and 34, though 19 SSB is said to be the toughest. And mind you, it is said that the Allahabad board is the toughest of all boards.'

On the next day, after settling in, the candidates faced

the Psychology Day. Jeetu goes on to write: 'We woke up at 4 a.m. and had our tea. You know we used to get up at nearly 4 a.m. every morning until our results for the SSB were announced and then we used to report at around 6.30 a.m. near our 19 SSB. The sun at that time was just rising and I used to think what kind of place have I landed in. You know quite well my sleeping routine but it didn't have any side effects on me.'

That was so typical of Jeetu, adaptable and ready to take everything in his stride.

On the same day, after a battery of psychology tests, they had the Thematic Appreciation Test (TAT), which was followed by the Word Association Test (WAT), and the Situation Reaction Test (SRT). Jeetu's interview too was on the same day, which lasted for half an hour and which he found 'quite exciting and satisfying'. He goes on to list all the questions that were posed to him and then writes: 'I will tell you the answers when you come to Udhampur.' Alas, I do not remember whether we ever discussed the answers.

The activities planned for the first day were over by lunchtime and then ten of them had gone to the Sangam, where the Ganga and the Yamuna meet. 'We reached the spot in about an hour. We went in a boat and sang, exchanged jokes etc, etc. The scene was really breathtaking, the blue waters of the Yamuna, the dirty waters of the Ganga and lots of river birds—numbering in hundreds, flying over our heads. It was a great experience,' he writes in his letter.

They had the Group Testing Officer (GTO) tests the next day. First was the group discussion, which had interesting topics like 'Should the army be allowed to maintain law and order and curb internal disturbances?' and 'How to battle the Kashmir situation?' These were followed by military planning.

'We had our command task, Progressive Group Task (PGT), Half-Group Task (HGT), individual obstacles, lecture (arms race), etc. With this my four days came to an end. On the fifth day, the result was announced, Naveen and I got through. Ten guys got selected from a batch of thirty-six. Then we had our medical and I came home yesterday. I haven't written this page in detail otherwise my letter will go to ten pages and won't fit in the envelope,' concludes Jeetu. Naveen Suri was Jeetu's classmate in APS, Udhampur, and is now a colonel in the army, currently posted at Udhampur.

Jeetu also described the food served at the SSB in detail: 'The food served by SSB was excellent. Being non-veg we used to get two boiled eggs, three slices of bread, butter, jam, tea, as much as you want, and custard. And at lunch we used to have rice, chapatti, one vegetable, dal, chicken or meat, alternately, and curd. At dinner we had a combination of lunch and breakfast. So, the SSB used to leave no stone unturned so that we get nutritious and abundant food...'

As an aside, at the end of the letter, in a box he mentions, 'I have got OIR 1 that is Officer Intelligence

Ranking 1, second in my batch.' On the margin of the last page of the letter, he adds a postscript, 'While writing this letter, I am half awake, exhausted after the tests and two days of continuous journey. I think I have made a lot of mistakes while writing the letter and my handwriting is also bad. After all, I am unique.'

Reading that letter after so many years, I am amazed at the way he has described the whole rigorous selection process, his keen observations and emotions so as to bring them alive for me. While reading it today, I could almost transpose myself to the SSB Allahabad and be there with him as he underwent all those tests. He was barely seventeen back then, but I found him to be extremely articulate in his written communication. All the reading that he had done all those years, helped, I guess.

Books were the favourite gifts for birthdays. Comics were also relished. Vinkoo recollects, 'Didi and Jeetu were Asterix and Obelix fans, but my favourite was Tintin.' Some of those comics are still lying around at home including the *Archies*, the *Phantom* comics, and *Amar Chitra Katha*. We grew up on a diet of Enid Blyton, Nancy Drew, Hardy Boys, Robert Ludlum, and all the classics of the Brontë sisters, Jane Austen, Charles Dickens, Thomas Hardy, Somerset Maugham, abridged versions of Shakespeare—the list goes on and on. The Northern HQ library was better stocked than the school library, and during summer holidays we would trudge to the library every other day. Even the librarian used to be amazed at the speed with which we would devour the books and come back for more.

All in the family were voracious readers. All volumes of the *Complete Works of Swami Vivekananda* by Sister Nivedita bought by Papa are still at home along with *An Autobiography* by Jawaharlal Nehru. Papa was quite fascinated with the inclusive conception of religion propounded by Swami Vivekananda at the 1893 World Parliament of Religions in Chicago, including his views on nationalism and universalism and on the detachment and the self. We were exposed to different streams of thoughts and were taught to respect other viewpoints.

However, as Jeetu started growing up, he became more focused in his reading. 'Right from childhood, Jeetu gravitated towards matters of security, both national and international, defence systems, anti-terrorism etc. I remember him reading a book on the Afghanistan crisis and the creation of the Taliban while he was in the sixth standard. At home, we discussed and debated a great deal about national and international matters,' remembers my sister. I too remember him getting obsessed with Afghanistan at one point. He had become a walking encyclopedia on the topic, including the three Anglo-Afghan wars.

One of his colleagues from his 34 RR days, Lieutenant Colonel Alok Dass, who has now hung up his uniform and imparts corporate training, very clearly remembers Jeetu's habit of reading. 'I am also in the habit of reading, but Shatrujeet read more than I did and was an intense reader. He had eclectic tastes and read books not only on military strategy but also on a wide range of topics.

'Shatrujeet would interpret a chapter, correlate it with data and while recommending books for me to read, he would specifically ask me to focus on some chapters. We would discuss books a lot and then I had this "silent pressure" to read before I met him again,' Dass recalls with a chuckle.

The Shadow of the Great Game: The Untold Story of India's Partition by Narendra Singh Sarila was gifted to me by Jeetu and Abha on 15 May 2007. They bought the book at a bookstore in the PVR, Priya Complex in New Delhi, and my copy has a short note scrawled on the second page in Jeetu's squiggly handwriting:

> Dear Chucks, I feel this book is a collector's item. I hope you will enjoy reading it as much as I did & you will come to know the intrigues of our partition—I am serious!!! So take out time and get going.

It was an order.

Jeetu was a leftie, and Lieutenant Colonel Dass also remembers how Jeetu would write smartly with his left hand: 'He did not have a great handwriting but it was clear and precise. The most interesting part was to watch him write with his left hand with an intent expression on his face.'

There were no mobile phones or internet to distract

us back then. Even the good old Doordarshan (DD) used to have patchy reception: the terrestrial antenna on the roof had to align with cosmic forces for us to be able to watch TV. It was Jeetu who would be on the terrace figuring out the direction, while one of us would shout from inside through the window: 'It—the picture—has come. Oh no! Gone again. Try the other direction.' This hide-and-seek would continue.

We were made to watch the 9 p.m. DD English news regularly to improve our vocabulary and diction and the weekly current affairs programme *The World This Week*. Watching *Chitrahaar*, featuring curated songs from Hindi movies, on Wednesdays and Fridays was totally optional! There was no bar on watching quiz shows and we would wait for Siddharth Basu's weekly quiz show, keeping our fingers crossed that both the power supply and the TV reception were in perfect alignment.

I think the 1970s generation has witnessed technological upheavals at an unprecedented scale. We have time-travelled from the audio cassette tapes to online music, having seen floppy diskettes, or floppies as they were called, and Compact Discs (CDs) on the way. From waiting in queues for making a call from the STD booth, late at night as rates were discounted, to the ease with which we use smartphones today, we have seen it all and embraced technology though I yearn for those mobile-free days where we were not at the beck and call of the world all the time.

Jeetu was a technology aficionado and loved

experimenting with whatever new technology had to offer in terms of products and services. He presented Mom her first Nokia mobile phone and patiently explained to her its features. Orkut, the social networking service, was launched by Google in India before the launch of Facebook in 2006. Jeetu was quick to join Orkut and tried unsuccessfully to persuade me to join as well. I am a misfit in this hyperactive social-media era and rather than scrolling through Instagram or posting on Facebook would prefer reading a book or going for a walk.

Colonel Harjinder Singh, Jeetu's Commanding Officer (CO) in the 34 RR, also remembers Jeetu's keenness towards exploring new technology: 'Shatrujeet loved to explore technology and he taught me how to transfer documents using bluetooth on mobiles way back in 2006-07. I was very impressed.'

The *Ramayana* and *Mahabharata* serials had the whole nation hooked, with roads empty during their telecasts. Ours was one of the few homes that had a TV, and half the village used to descend on our house to watch these serials. The three of us would go and sit in the garden because there would be no place inside. I think these serials were a turning point for DD.

In 1989, Shahrukh Khan made his debut on Indian TV with the serial *Fauji*, which followed the trials and tribulations of a young commando officer. Mom was extremely fond of the show and used to say, 'This boy is very good. Mark my words, he will go far.' I guess her prophecy has come true. I believe, we have another connection with Shahrukh—according to me, there is some physical resemblance between the younger Shahrukh and Jeetu, including their mischievous, impish smile.

Music was another constant in our growing-up years. 'Limited by the scope of accessing content during the 1980s, we grew up on a diet of reading and listening to music on the radio and a Bush stereo system,' remembers Vinkoo. The Bush stereo system was one that Papa had bought from the army canteen. Though Udhampur is a small town, as a cantonment city, it had shops selling the latest English and Hindi audio cassettes, catering to all tastes. Initially, Papa bought cassettes of Ghulam Ali's ghazals, Anup Jalota's bhajans (for Mom), and Boney M and Carpenters (for the three of us). Later on, the three of us would save money, even single rupee, throughout the year and dispatch Jeetu to buy cassettes of Grammy winners from the music shop in Chopra market. It was economical to buy just one cassette of Grammy winners as one could hear the best of many singers. We would keep playing the cassette till it groaned and gave up. I still remember Tracy Chapman winning the 'Best New Artist' at the 1989 Grammy awards for her song 'Fast Car'.

My sister remembers: 'Jeetu was a fan of English pop music and English movies. Among the bands or singers that he followed were Boyzone, Bryan Adams, Bon Jovi, Queen and MLTR. He used to sing a few songs very well, like "18 till I die", "Summer of '69" and "It's my life".

In school, he used to maintain a diary, write down the lyrics of the songs that he liked and sing them whenever he wanted to.'

Jeetu's CDs are still lying around at home. These are recorded CDs with songs of Savage Gardens, Metallica, U2, Ugly Kid Joe, Vanilla Ice...and I was quite surprised to find one of Mirza Ghalib in the collection. Arjun remembers: 'During one of his annual leaves, Jeetu Bhaiya had replaced the old TV and also bought a five-in-one. Bhaiya used to bring movie CDs from the shop in Chopra market and we would watch them the whole night. *Top Gun* was his favourite movie and I think we watched it twice or thrice.'

Though we were to hone our English reading as well as speaking skills, we spoke Hindi, English, and Bhadarwahi at home. Bhadarwahi is an Indo-Aryan language of the Western Pahari group, and was documented in Sir George Grierson's *Linguistic Survey of India,* Vol. VIII, Part I in 1916 as one of the dialects of Kashmiri. Apart from my mother, I am the only one in my family who can now speak it. I make it a point to converse with Mom in my rusty Bhadarwahi, but none of the next generation know the language. Even my sister, as a child, never spoke it though Jeetu used to make an effort. I used to find these attempts of his very comical, and would tease the poor chap when he tried.

Mom is a treasure trove of Bhadarwahi proverbs: she pulls one out of the hat for every occasion. This is linguistic history too—this almost forgotten collection of proverbs—and I want to document them. A 2022-report of the World Economic Forum notes that 1,500 known languages may no longer be spoken at the end of the century, and I fear that Bhadarwahi too may share the same fate. Loss of a language is a collective loss of a culture, impacting our identity and belongingness.

Apart from studies, we were encouraged to participate in extracurricular activities like debating, declamation and quiz competitions. Preparing for debates used to be a family affair with all of us pitching in to prepare the text and then holding mock sessions. 'Shatrujeet was a good debater and had a great command over the English language. He would also participate in plays and loved to act,' recalls Ranjeet. Jeetu's panache for acting came as a surprise to me; I had not been aware of his acting prowess.

All three of us won many prizes in debates and other competitions. The shields and cups we won are still in the Udhampur home and Mom keeps getting them cleaned and polished. Jeetu was particularly good at quizzes. He was part of his school quiz team and I think when he was in the seventh or eighth standard, his school team won the inter-district competition held in Jammu. Apart from a shield, each member of the team was also given a steel bowl—Jeetu's is still used at home.

There was a rule at home—we had to eat whatever was cooked and served. Vinkoo also recalls: 'As a child, Jeetu used to eat everything that was served at home except for the greens. He used to keep that portion for the end, close his eyes and swallow them fearing a reprimand from Mom, who always ensured that we ate enough greens. He was also extremely fond of the Brown & Polson custard. I still remember Mom making custard daily for him.'

Vinkoo then continues with a laugh, 'He was also very fond of besan ladoos. I remember making ladoos for him whenever he would return to the NDA or IMA after a break or a holiday.'

This reminds me of an incident. Uncle Devraj had a grocery shop at the end of the kaccha road, and his was the only grocery shop in the vicinity then. All provisions for the month were ordered from there, and the account would be settled at the end of the month. One time, the bill was on the higher side, and when Papa delved deeper into it, he found that the amount included a charge for around seventy or eighty besan ladoos. There could be only one suspect, and Jeetu very sheepishly admitted to eating them over the whole month. All of us had a good laugh, with poor Jeetu blushing scarlet like a beetroot.

Dinnertime conversations would revolve on a multitude of subjects. Open discussions were the norm, and we were encouraged to air our views. Sometimes there would be heated arguments, and Papa would advise, 'Do not be afraid to express your views on a subject. However, be clear with the facts, cogent in your articulation and do not argue only for the sake of arguing.'

I think this advice has stood all of us in good stead.

Colonel Hasabnis, Jeetu's CO in 2000 during Jeetu's posting at Uri, recalls: 'Shatrujeet was my adjutant and played a very crucial role in keeping communication channels open with officers as well as troops. However, what I remember most clearly was that he had his own mind and would not hesitate to offer suggestions. In the army, an order is an order but he would temper it with his thought process and present alternatives too.'

Not doing well in studies, whether in school or otherwise, was not an option—all three of us had to do well. Mom's favourite quote is: 'There are no shortcuts to success. The road to success is paved with sweat and toil. You need to have the grit and determination to succeed.' Our parents came from the old school of parenting in which 'hard work never kills anyone' was the mantra, though Papa would be more indulgent. Today, Mom sometimes feels we are too soft with our children but, so as not to ruffle any feathers, strategically refrains from commenting on our parenting styles.

Ours was a close-knit family with strong connections with the extended families on both sides, Mom's as well as Papa's. Like migratory birds, winters would be the time for Daadi and other relatives to descend on our home as it would snow heavily in Bhadarwah and get too cold for Daadi. My buas (paternal aunts), mamu (maternal uncle) and their families would come visiting as the schools were closed in Bhadarwah for the winter holidays. Schools would reopen in March. However, for us it used to be examination time. I can't fathom now how we studied amongst all the hustle and bustle! On top of it, Mom would manage her office work, guests at home; supervise her three kids plus Arjun and all the cooking without a frown on her brow. Now, the mere thought of a few guests coming over throws one into a tizzy.

The cherry on the cake would be the Sunday breakfast-eating competitions between Jeetu and Anchal, our mamu's son, who is now a urologist. They would be competing on who could eat more idlis or khameeras, the thick fried rotis made with yeasty wheat flour dough. Everyone would gather around them, cheering the two gladiators in the arena while Mom anxiously eyed the remaining idli batter or the dough. She had others to feed too. What a sight it used to be, watching them wolfing down fifteen-to-twenty idlis.

The Bhagavad Gita records the conversation between Krishna and Arjuna right before the epic battle of Kurukshetra. At home, there used to be a poster depicting this famous scene in the Mahabharata, hanging below a wall clock with following text (verse 2.47):

> कर्मण्येवाधिकारस्ते मा फलेषु कदाचन।
> माकर्मफलहेतु भू:मातेसङ्ग अस्त्वकर्मणि॥
>
> Karmany e Vadhikaras te Ma Phelsu Kadacana
> Ma karma-phala-hetur Bhur ma te Sango 'stv akarmani

Loosely translated, it means, 'You have a right to perform your prescribed duty, but you are not entitled to the fruits of action. Never consider yourself the cause of the results of your activities, and never be attached to not doing your duty.'

Papa would keep quoting from the Gita, and this was his favourite verse. The three of us would roll our eyes in exasperation and try to distract him before he got going full steam!

Be honest to yourself and your work. Like a woodpecker pecking on wood, our parents drilled the values of hard work, honesty, integrity and empathy into our heads. At the drop of a hat, Mom would quote this couplet by Tulsidas:

Daya dharam ka mool hai, paap mool abhimaan,
Tulsi daya na chodiye, jab lag ghat mein praan.

Kindness is the root of the religion and pride
is the root of sin;
Tulsi does not let go of kindness even when
life is in danger.

I think her grandchildren—Shravani; Anadi, Vinkoo's son; and Kartikeya—too have heard it umpteen times!

Mere words can be hollow—we teach our children by our actions. I have watched Papa take off his pullover and give it to a beggar standing at the gate outside our house shivering in the cold. I have seen Mom, with Papa in tow, carrying a torch, going up to the village in the middle of the night to give an injection to a patient. There would be many women visiting our house to get injections administered, and Mom did not charge a single penny from anyone.

Another thing that Papa was particular about was how we treat people, particularly those who work for us and our juniors. He had a simple logic: 'You in any case will be deferential to your seniors but how you treat your juniors is a reflection of your character.' We were taught to treat everyone with respect irrespective of their station in life. No driver could leave our house without having water or tea. Even the orderlies that worked for Papa were treated like part of the family. There is one picture of the three of us with Suryakant Bhaiya, Papa's orderly in Guwahati. Jeetu, barely a year old, is sitting on his lap. Shravani has a good laugh looking at the picture of the four of us with somber expressions on our faces and particularly my oversized 'retro glasses,' as Shravani remarks. Suryakant Bhaiya was a gruff man and a disciplinarian with a soft heart. He was from Gujarat and would bring groundnuts for us when he came back from his annual leave.

Ami Lal, Jeetu's orderly in Pune, says, '*Sahib ke bare mein jitna bole, utna kam hai* (No matter how much we speak about him, it will not be enough). He was so good with everyone *aur ab sirf yaadein reh gayi hai* (only memories are left now)'. Ranbir Singh, Jeetu's orderly-cum-Man Friday who was with him right from 1999 till June 2005, remembers with choked voice, '*Sahib se itna prem, koi rishta purane janam ka tha* (Had so much of love for him, it must have been an old relation from previous birth).'

As they say, a person is the sum of many parts. There is much more to life than passing exams, landing a well-paid job, having a palatial house and a big SUV—the barometer on which we judge people now. What matters is the entire cultural package of an individual. Our parents provided this whole cultural package of a world full of books, movies, music, ideas and values to us in our growing-up years. This has held us in good stead as the three of us embarked on our different career paths.

3

Journey Through the NDA and IMA

Searching in Journals, Letters and Conversations

Some of them take root in my soul,
And like a possessed wanderer,
I embark on the journey,
To search for their origin.

'Journaling is mindfulness in motion', says Lisann Valentin, a shamanic life coach. I have kept journals since my school days, long before journaling became the buzzword for mental wellness. Scribbles in these diaries and notebooks are irregular, random, generally recording my observations on events happening around, and my reactions to them. Most of the journals have since been lost and only a few remain—and even those end abruptly. While going through the remaining ones, there is one entry dated 10 January 2005:

> I woke up disoriented with a strange feeling. I had a bad dream. In the dream, I saw that Jeetu had died and there

> were strange people in our Udhampur house. I cannot see any familiar faces like my Mom, Madhu or Shravani.

I had forgotten about this entry and chanced upon it only while foraging for material for the book in my journals. Did I have a premonition? Does the unconscious know what is in store for us? Today it reminds me of Joan Didion's fascination with what she once called 'the unspeakable peril of the everyday', which I would get familiar with while reading *The Year of Magical Thinking*.

The last entry in that journal is dated 4 April 2008, and on 17 June of the same year, barely two months later, our brother passed away. On one of the pages before the last entry, I have pasted the poem 'Song of an old woman abandoned by her tribe'. It is an anonymous southern Sohone song that has been reworked by Mary Austin, and part of it reads as follows:

> Alas, that I should die,
> That I should die now,
> I who know so much!

Reading it now, I am struck by the thought that it could be my brother's lament too. He was wise beyond his age. In one of the courses he did at the College of Military Engineering in March-April 2005, he got an overall grading of 'A', and his pen picture reads as follows:

> A smart, intelligent officer who came adequately prepared for the course. He is very resourceful and

> decisive, and has a forceful verbal expression. He is mature and exerts a good influence over the group. He has a very sound practical sense and remains unperturbed under stress. He thinks logically and arrives at workable solutions to practical problems.

My first thought on reading this entry was that the officer who wrote this pen picture was quite intuitive and had captured the essence of Jeetu's personality. Jeetu was Zen-like in real life too: calm, composed, and unhurried; not the one to get perturbed easily. It used to frustrate Mom no end that he kept things pending till the last minute and was laid back. This was her constant refrain against him since school days. But beneath that calm veneer was a mind that was disciplined, logical and meticulous in his work. While leafing through Jeetu's small scribbling pad, I realized how meticulous he was. There are 'to do' lists and those include tasks like fixing the Velcro on his cummerbund, little notes to self, and minute observations on tasks at hand. He had even recorded the names of movies that he watched in the NDA during February 2008!

Colonel Aditya Verma, an artillery officer who is now a Deputy Commander and posted in Baramulla, Jammu and Kashmir, was Jeetu's colleague in the 34 RR. He recalls: 'Once Shatrujeet was on leave and I was officiating in his place. I visited his company HQ and there I found a register maintained by him. It was meticulously prepared with complete details of all the villages under his responsibility, including names and contact details of the prominent people of those villages. I found this to be a very good practice. I implemented it in my company too for my area of operation.'

This reminded me of a quote by Michael Caine: 'Be like a duck. Calm on the surface, but always peddling like dickens underneath.' Jeetu was like that duck, constantly striving for excellence without making a hullabaloo about it. He marched to his own tune, strove for excellence, and set his own goals.

We have grown up on Linda Goodman's *Sun Signs* and although it is a cliché, Jeetu was a typical Leo. A natural leader, courageous, and believing in living life to the fullest. 'It's my Life' by Bon Jovi was one of his favourite songs.

It's my life
It's now or never
But, I ain't gonna live forever
I just want to live while I am alive.
...
Better stand tall when they're calling you out
Don't bend, don't break, baby don't back down
It's my life.

Maybe this song reflected his philosophy towards life—grab it by the collar and live it to the fullest, even though it may be short-lived.

He was not the one to blow his trumpet or be envious of others. This is not to say that he was not competitive or ambitious; he was, and in the same long letter dated

26 February 1995 that he wrote after coming back from the SSB, Allahabad, he wrote the following in bold letters:

I HAVE QUALIFIED MY SSB AND MEDICAL

It was the cry of exultation of a seventeen-and-a-half-year-old boy on cracking the tough exam and the even tougher SSB that enabled him to enter the hallowed portals of the NDA, Khadakwasla, Pune.

The idea of having an institution like the NDA was conceived after the Second World War, when it was recognized that a 'jointness' in modern warfare was required in the country. The foundation stone for the permanent war academy at Khadakwasla, was laid by Prime Minister Pandit Jawaharlal Nehru himself on 6 October 1949. The academy was formally inaugurated on 16 January 1955 and since then, it has been the training ground for cadets of the three wings of the armed forces, who train together for three years. After graduating from the NDA, they go to their respective specialized training academies. Its central façade is a true Joint Service Design: the majestic Sudan Block resembles a field gun of the army; on the right, the Vyas library and the Naval training team, which resemble an anchor; and on the left, the Habibullah Hall that resembles an aircraft.

Interestingly, the Government of Sudan in 1941 had given a generous gift of a hundred thousand pounds to Lord Linlithgow, the then Viceroy of India, to build a war memorial to commemorate the loss of lives of Indian troops during the liberation of Sudan in the Second World War. It was this core corpus that was used to build the Sudan Block, the grand edifice of the NDA.

Habibullah Hall seems to be quite a popular place! In his letter dated 5 October 1995, written around four months after joining the NDA, my movie-buff brother writes:

> I went to Pune, saw *Rangeela*. It is nice. Have you seen it? During the midterm, I saw four more movies at our Habibullah Hall, namely *Aasmayish, Taqderwala, Krantiveer* and *Vijeta*. *Vijeta* is a movie about the NDA and is made by Shashi Kapoor with his son as an NDA cadet.

Twenty-eight years later on 21 June 2023, to celebrate the silver jubilee of their batch, I was sitting in the same hall with Vinkoo, Abha, Kartikeya and Jeetu's batchmates and their families, watching a video titled the 'Fallen Heroes: 94th NDA Course'. They have lost twenty-one of their batchmates, some as young as Flying Officer Chaitanya Marathe who was twenty-two when he died in an aircraft crash in October 1999, barely four months after passing out from the IMA. Chaitanya's mother and sister were sitting next to me and in silence we commiserated with each other on our shared loss.

Major Sandeep Unnikrishnan, who was killed in action during the November 2008 Mumbai terror attack was also Jeetu's NDA batchmate. He was awarded the Ashok

Chakra (AC), the highest peacetime gallantry award in January 2009. He too was born in 1977 and died barely five months after Jeetu's death. A film titled *Major* has been made on his life and was released in 2022. Sandeep's parents had also come for the silver jubilee function.

Major K.P. Vinay, fellow NDA batchmate, was Jeetu's colleague in the 34 RR during 2006-2007 in the Kashmir valley. He also died fighting the terrorists in October 2007 and was awarded the Kirti Chakra. We met Vinay's mom, brother and his family there.

There were relatives of other fallen heroes too—parents, siblings, spouses but not all had come. I could understand their absence too. It is oft repeated that time is a great healer. Is it really? Does the gnawing feeling at the pit of your stomach go away ever? Do your eyes not well and your throat not choke looking at the old photographs? Pico Iyer in his book *Autumn Light: Season of Fire and Farewells* recounts that Henry David Thoreau, the American poet, naturalist and philosopher, till the end of his days had to leave the room whenever his late elder brother John's name was mentioned. John was only twenty-six when he died of tetanus in his younger brother's arms.

In his letter dated 5 October 1995, Jeetu goes on to write,

> OTA, IMA and INS Shivaji cadets have come for seven days to participate in games across various categories. In midterm, I volunteered for a cycling hike to Bhaneshwar, which is forty kilometres from the NDA, so the entire

> trip, to and back makes it eighty kilometres. Basically, I wanted to check if my stamina could handle the distance. We started at about seven in the morning with our packed lunch and breakfast. On the way there was a six-kilometre-stretch, which was all uphill. After coming back from Bhaneshwar, we were taken to SNAKE PARK, Pune. It was quite eerie with wells full of snakes... By the time we came back we had stiff legs and cramps. But, as a whole, we enjoyed it.

It was fun mixed with loads of hard work.

Jeetu was a born runner and he writes,

> We had our novice cross country also. You will be happy to know that out of 340 cadets, I came seventh running a distance of 7.15 km in 22.06 minutes. The first cadet was from RIMC, he clocked 21.28 minutes.

In his letter dated 2 August 1996, he writes,

> Starting from this term, we have X country and we used to go for runs nearly every day. Running long distance really takes you up. It ended on 28th of July and I came in the third enclosure. There are ten enclosures. You know after a long time, the whole academy was running and nearly 1,500 cadets were running. It is a huge mass and when it starts there is a virtual stampede...

Even during our conversations, I remember discussing with Jeetu the joys of running. He liked the focus and, I guess, the solitude it offered. At the NDA, they were made to run fourteen kilometres every day in the third semester and in one of the letters he complains about shin pain, cramps, and knee issues. Then very courageously he writes, 'But, don't worry I am alright and I will take care of myself.'

Dheeraj recalls, 'We ran many cross-country runs both in the NDA as well as the IMA and ended up next to each other. As we belonged to different squadrons, we were expected to compete with each other and overtake each other. But we ran in pairs and were known as the Kotwals, and people thought we were cousins.' He chuckles softly at the memory.

Haruki Murakami, the world-renowned Japanese writer, began running in 1983, at the age of thirty-four, just to keep fit. This was one year after he had devoted himself totally to becoming a novelist. However, while reading his book *What I Talk About When I Talk About Running,* I realized running was more than a physical activity for him. He writes, 'I know that if I hadn't become a long-distance runner when I became a novelist, my work would have been vastly different. How different? Hard to say. But something would have definitely been different.' This meditative memoir of running, writing peppered with Murakami's music choices, would have appealed to my brother.

While discussing the qualities that a novelist must possess, Murakami lists three—talent, focus and endurance—with the emphasis being on the last two. I think these three qualities are equally relevant in any role that we perform. Murakami also does not gloss over the toll that running takes on your body, but I guess it becomes addictive after some time. Dheeraj has continued with his long-distance running and he tells me, 'Running is not only about lung capacity; it is about determination and perseverance.'

While running, I use music as a prop—with music streaming into your ears, feeling your own breath, the rhythmic thumping of the feet on the ground, you are one with your consciousness and it can be cathartic.

Our handsome brother, who got the title of 'One with most coveted eyes' after his 12th standard, was wary of taking up boxing at the NDA as he didn't want to get his face broken. He writes, 'You get your face only once in a lifetime! Isn't it?' Yes, brother, how right you were.

For all the youngsters joining the NDA, life is not easy, especially during the first two terms. It can be physically exhausting and mentally challenging. In a letter written during the first semester, Jeetu writes, 'Our midterm has ended and now regular classes have started. You know there are just fifty-seven DLTGH. When I came, it was 154.'

Later on, I learnt that DLTGH stands for 'Days Left To Go Home' and I understand that every ex-NDA cadet remembers it. Going home was liberating for each one of them.

In his letter dated 12 January 1996 Jeetu writes,

> I enjoyed my holidays and met my friends. Udhampur is the same old place, which is quite slow. I saw DDLJ in Delhi while going back to the NDA. You know I had complete rest for about thirty days...I used to sleep till ten in the morning and then roam about. I saw ten movies, went for a picnic to Jhajjar Kotli, and then attended two get-together parties. You know, Mom used to say that when she goes to the office I am sleeping and when she comes back, I am outside.

Vinkoo recalls: 'From the very beginning, I never heard him cribbing. He always took everything in his stride, never getting bogged down by any pressures of life. During his entire training at the NDA or later at the IMA, I never heard him crib or complain about the rigorous training that every cadet undergoes. I still remember, when he came back home after the first break from the NDA, his hands were full of blisters. I looked at him and he just smiled—that beautiful smile of his that always reached his angelic hazel eyes. Never did he complain about the tough and rigorous training that makes a boy a gentleman cadet. All throughout his life, he never ranted, cribbed, or complained about anything in life.'

Dheeraj recalls his first impression on meeting Jeetu. 'I had joined the NDA in August 1995, while others were already there since July 1995. I was coming from Sainik School and when I met him for the first time, I guess we were sizing each other up.

'I thought he was too sophisticated, too polished and may not be physically fit. However, I was proven wrong,' remembers Dheeraj wryly.

Jeetu had a lean, slight frame but he was extremely good at sports. As Dheeraj recalls, 'Shatrujeet was a leftie and it gave him an innate advantage in games. He was a very good handball player. However, he loved swimming and was an excellent swimmer and a super water polo player. He won many medals for the battalion team.' I think the swimming during summer breaks while at home had made him a good swimmer. The medals displayed in a glass showcase on a wall in our Udhampur home are a testimony to the fact.

Of people no longer there, the memories get embedded in inanimate things: I still have Jeetu's lemon-yellow NDA T-shirt that he gave to me and the parrot-green bandhani dupatta that he and Abha got from Rajasthan where they had gone for their honeymoon in March 2004. Vinkoo also has a kiddie T-shirt gifted to three-year-old Anadi by Jeetu and Abha in 2007 with 'Life is Good' emblazoned across it.

Our father died of cardiac arrest in his mid-fifties in February 1996. Jeetu was not even nineteen and was in his second term at the NDA. With misty eyes, Vinkoo remembers: 'By the time Jeetu reached Udhampur, Papa had been cremated. I do not recall Jeetu crying. During the day, he would be so calm and serious, helping around, talking to relatives. However, at night he would come to where I was sleeping, lie next to me, saying nothing, and just hold my hand. We were trying to take strength from each other and grieving quietly in the stillness of the night.'

Dheeraj recalls, 'We came to know that a cadet had left for home on emergency leave. It was only later after Shatrujeet had returned that I realized it was him.

'Losing your father is such an emotional setback; however, Shatrujeet handled it with so much poise and courage. He was calm and collected,' remembers Dheeraj with admiration. Similar thoughts are echoed by Ranjeet: 'He lost his father while so young but he handled this adversity with strength and fortitude. But, I think this hastened his growing up from a boy to a man. He felt responsible for the well-being of his mother and sisters.'

The three-year training at the NDA chisels you. All these youngsters joining at the age of seventeen or eighteen are put through rigorous academic, physical and service training. They are awarded a bachelor's degree by the Jawaharlal Nehru University (JNU), one of the premier universities of the country. Jeetu graduated with a Bachelor of Arts degree in history, economics and geography. Jeetu had also taken German as a foreign language and a few of his letters were peppered with German phrases duly translated. In one letter, he has signed off with flourish—'Your Bruder'; Bruder in German means brother.

The focus in the NDA is on the overall personality development of the cadets to instill in them the highest values of integrity, honour, self-discipline, and camaraderie.

Esprit de corps is born there. They are bound by their sweat, toil, collective punishments and laughter for the rest of their life.

I remember Jeetu talking about his batchmates with great fondness, particularly his squadron mates. He was in the Mike squadron in the NDA, and in Jeetu's papers I found a congratulatory letter dated 24 April 2008 from Major Milind Shinde, one of his juniors in the Mike squadron.

> Dear Maj Kotwal,
>
> I'm Mike-99, I was your 1st termer when you were SCC. I'm located at the moment in a post at 14,000 feet. It was in the Feb issue of *Baatcheet* that I saw your photograph and name and also that you have been awarded the Kirti Chakra. Sir, I take this opportunity to congratulate you for the award and wish you many more endeavours in the future also...

On Googling, I found that *Baatcheet* is a magazine brought out by the Ministry of Defence for dissemination of information. Here is a young officer sitting at a post high in the mountains, happy at his senior's success and reaching out to him. It is touching to read about these bonds forged and strengthened over the years.

While coordinating our participation at the silver jubilee celebrations, Colonel Aditya Kanwal reached out to me, courtesy Dheeraj again! Colonel Kanwal invited me to the silent wreath-laying ceremony at the National

War Memorial, Delhi. On the bright, sunny morning of 12 June 2023, I attended the ceremony that is always organized by Jeetu's batchmates on this day every year, the day they were commissioned to pay homage to their course brethren. The National War Memorial, inaugurated in 2019, consists of a central obelisk and four chakras or circles. At the heart is the Amar Chakra (Circle of Immortality) followed by the Tyag Chakra (Circle of Sacrifice), third is the Veerta Chakra (Circle of Bravery) and the last one, the Raksha Chakra (Circle of Protection). Colonel H.S. Lidder, as representative of their batch, laid the wreath at the base of the obelisk, below which lies the eternal flame, which is symbolic of the immortal spirit of fallen soldiers. It is an assurance that the Nation will never forget their sacrifices.

I met Jeetu's batchmates, some of whom were there with their families. It was surreal being part of the ceremony. I was meeting them for the first time as Shatrujeet's sister and there was some awkwardness. I was part of the ceremony and yet watching it from a distance; I was there but not part of their group. Then Jeetu's Mike squadron mates reached out to me and suddenly I felt a kinship with the five of them who were there. I was invited to their group WhatsApp call in the evening. Few of them reached out to me after the call. These men were truly a 'Band of Brothers'.

Getting assigned to a particular squadron for three years gives them a unique identity that binds them together for the rest of their life—whether in service or not. Mike squadron is referred to as the Jurassic Park by ex-NDA guys as this was part of the 4th Battalion—the last battalion to be raised when Jeetu was at the NDA. At that time, Mike, November and Oscar were the only three squadrons of the 4th Battalion. Now, I gather a 5th Battalion has been raised in 2012 and three more squadrons added, increasing the total number of squadrons in the NDA to eighteen.

All of his batchmates are in their mid-forties now and looking at them I was trying to visualize how Jeetu would have looked. Grey hair on his temples, his handsome, rugged face with laugh lines around his hazel eyes, ramrod straight without an ounce of belly fat and medals glistening on his chest...I am sure he would have remained physically fit like some his batchmates. Being 'fighting fit' was always important to him. He, however, never got to be old and remains 'forever young' like Neil Young's song:

> May your song always be sung,
> May you stay forever young,
> Forever young, forever young.

On 16 June 2023, I met Colonel Kshitij Balhara whom I had met briefly at the silent wreath-laying ceremony. Kshitij is a sensitive, reticent man, quiet in a brooding sort of way. Our conversation revolved around Jeetu and I could sense the deep feeling of grief that he still carries within him about Jeetu's death. 'I was posted as aide-de-camp (ADC) to the Central Commander then and we were to visit the NDA in mid-June. I had spoken to Shatrujeet a few days back and we had made plans to meet. I just could not make myself go to the NDA after his death,' recalls Ksihitj with a faraway look in his eyes.

Kshitij also recalled with clarity the help Jeetu would extend to him and other batchmates. 'We bonded well as both of us came from an army background and were neither from a Sainik School nor from the Rashtriya Indian Military College (RIMC), Dehradun. Guys from these two institutions have an inherent advantage in terms of their physical prowess. I was not very good in physical tests and had a tough time reaching the benchmarks. Shatrujeet cleared his physicals with ease, as good as cadets from those institutions. He was extremely good at swimming and long-distance running. He would keep pushing and motivating me during my physical exercises. On certain occasions there was no need for him to go swimming but he would cycle from the squadron to the swimming pool and be there to extend a helping hand to some batchmate or the other who was struggling with his swimming.

'Shatrujeet would keep telling me there is no need to compete with these Sainik School and RIMC types, just be clear about what you can do. He was not trying to ape anyone but rather was focused on improving himself. I have not met anyone so young who was so clear in his head,' reminisced Kshitij.

According to him, Jeetu was a perfect triangle of: Ability, Capability and Ambition. 'Shatrujeet was aware of his capabilities and abilities and the ambitions were tempered accordingly.

'He was sure about himself and was not competing with anyone. Whenever, I would get agitated about some test result or something, in his patent laconic manner he would say, "Chill kar Balhara",' recalls Kshitij.

On a rainy evening on 4 August 2023 in Delhi, I met five of Jeetu's Mike squadron mates from the NDA, all Colonels now—Ashwini Mishra, Arun Nayan, Biswajit Dash, Sharat Kumar Mishra, and Kshitij. The meeting was coordinated by Kshitij. Meeting them reminded me of the oft-quoted couplet of Urdu poet Majrooh Sultanpuri's poem: '*Main akela hi chala tha janib-e-manzil magar / Log saath aate gaye aur karwaan banta gaya* (I had started towards my destination alone but / People kept joing me and a caravan was formed).'

Much like the poet, I too had started almost alone on this journey to chronicle Jeetu's story. But, as I trudged on, others came forward and joined me and our procession kept growing. I was no longer alone.

The first to start talking was Biswajit Dash. In the last term at the NDA, Biswajit was made the Squadron Cadet Captain (SCC) of the Mike squadron, an honour bestowed on an all-rounder cadet. However, due to a friendly incident, which all of them narrated with great gusto, Biswajit was detabbed, and his SCC tab was taken away. But, Biswajit was asked to suggest his successor. 'Shatrujeet was my first choice as he represented the true spirit of the NDA. He was calm, composed, sincere and thoughtful. I knew he would be a balanced leader and I was not proven wrong. He did remarkably well as the SCC. He would come and discuss things with me though there was no need for him to do so,' recalls Biswajit.

Listening to Biswajit, I could connect the dots and finally understand what Major Shinde had meant when he was referring to Jeetu as the SCC. However, I do not remember Jeetu ever mentioning this to me.

'While playing football, I broke my thigh bone in the third term, was operated upon and had a rod inserted in my femur. I was hospitalized and hence, lost one term. Shatrujeet was one of the few cadets who would visit me in the hospital and provide that emotional support to overcome the difficult time. On getting discharged, I had this mental block and was not able to run. To make matters worse, I was feeling low as I had lost one term. Shatrujeet was there to help me overcome both physical and mental barriers. He trained with me rigorously and stood by me like a rock,' remembers Ashwini.

'Shatrujeet was also very candid to acknowledge his shortcoming of not doing too well academically in the initial terms,' remembers Biswajit. Vinkoo in one of her letters dated 2 January 1996 writes, 'Time flies very fast. Jeetu just came and left. He slept a lot during his holidays... He is faring well at the NDA. He lost the Bronze torch by a few points, but has become the understudy of the Squadron Cadet Captain (who is a sixth termer).

Understudy is like a PA and quite an important post...'

Gold, Silver and Bronze torches were awarded to the cadets who were doing exceptionally well academically, and they would wear them proudly on their uniforms. It was probably his not great academic performance in his first term that cost Jeetu the Bronze torch.

While speaking to Jeetu's friends, I could feel the warmth and respect that they had for him. Two things stood out from these conversations. One, they were all unanimous in their awe of the fact that he ran the cross-country race in his sixth term and got a medal. Arun, the most exuberant of the group, remembers with a laugh, 'Who runs the cross-country race in their sixth term? It was such a strange decision. This is the term to chill, relax and take life easy, when you are at the top of the pecking order. But, Shatrujeet was an exception. He trained rigorously, participated and won a medal. What a feat!'

Second, they were all impressed with his principled approach to life and sense of righteousness. Sharat Mishra remembers, 'All of us were young, seventeen or twenty years old, but for one so young, Shatrujeet was very principled. He had this innate maturity to handle situations and people, and had a calming effect on those around him. No shortcuts for him to achieve things.' Echoing the same thought Arun also shares, 'The Mike squadron was then famous for its ragdas (daily grind), which usually featured us lying and rolling on the ground as a form of punishment. Each mate had to follow a specific speed in tune with the rest of the squad, lest they be "rolled over" by their other mates. During one of such after-dinner ragda, Shatrujeet was asked to "roll-over" his slower squadron mate. He could have easily done that and gone off to sleep in his cabin. But, he refused to.

'In the middle of a six-month-term, there's always a midterm break of three or four days when no routine is followed and cadets generally chill and relax. However, they cannot go home. During the midterm break in the third term, few of the cadets had sneaked off to Bombay. Kshitij and Shatrujeet bumped into them at the Pune railway station where they had gone to drink lassi but were not aware that these guys were going to Bombay. They came back and went to sleep. Next day all hell broke loose when, during roll-call, we realized that some of the cadets were missing. Shatrujeet and Kshitij put two and two together but kept quiet and bore the collective punishment with everyone else. They would not rat on their squadron mates and take the easy way out!'

However, no talk of the NDA days can be complete without touching upon the topic of food. All of them remembered being perpetually hungry. Often, one or the other would be late for breakfast or had to forgo it entirely as they were completing some type of punishment. Those breakfasting would then surreptitiously smuggle bread slathered with jam and butter or a boiled egg in their satchel for their cadet brother and give it to him during the classes that followed breakfast.

They also regaled me with their vivid descriptions of the special lunches on Sundays at their local market. First, they would gorge on chole-bhature and then polish off a brick of ice cream ringed with ten or more hot gulab jamuns. One bite of plain vanilla ice cream followed by a hot, juicy gulab jamun. Pure ecstasy! On either side of the plate would be a bottle of cold Thums Up to wash down the meal. Some of them tried to replicate this feat during their subsequent trips to the NDA but could not even complete half the meal. This was a feat that could be undertaken only by twenty-year-olds undergoing rigorous physical activities coupled with high metabolic rates.

They went into raptures describing the pink and white doughnuts that would be available in the bakery in Gol Market within the NDA campus. However, money was always in short supply as they were allowed a paltry monthly allowance, which was paid on payday by the Cadet Quartermaster and promptly spent on food. Arun remembers using the hot water from the geyser in the bathroom or an immersion rod to make Maggi noodles. Once Arun had cooked thirty-six packets of Maggi with ten eggs topped with dollops of desi ghee that he had brought from home. But, the best was using their 40-volt table lamp to cook the instant noodles. Before leaving for one or another physical activity, Maggi mixed in water and masala would be kept under the lamp, which would be covered with their sweaty olive-green jerseys. By the time they came back, it would be cooked. They even ate the raw Maggi noodles, such were the pangs of hunger. Listening to all these Maggi tales, I was wondered if Nestle had ever covered the NDA in its sales market survey? The Academy would have topped the charts. Other than relying on instant noodles, they had lucky days with promises of samosas and jalebis. These were usually birthday party staples.

Their talks of NDA days were tinged with nostalgia. It was a brotherhood earned through trust—living through tough times together and the certainty that they would always have each other's back. They were also very proud of the fact that their squadron was the most decorated squadron of the 94th batch with the most gallantry awards—one Kirti Chakra, four Shauraya Chakras and three Sena medals.

During their training, cadets also form strong bonds with the people who train them and serve them. All of them remembered Kambli, a civilian employee who would get their clothes from the dhobi (washer man) and do sundry chores for them. Arun had met the drill master at the twenty-fifth batch reunion in June 2023 and was amazed to know how he still remembered his name.

Jeetu's squadron mates were unequivocal in their opinion that he was destined for a career in the services. According to them, Jeetu had the qualities of a good leader—honesty, empathy, reliability, sound judgment and most importantly, an underlying moral compass. There was no vengeful bone in his body; he had a constructive approach to do things rather than blindly following traditions.

While speaking to his batchmates, I recalled his 2 August 1996 letter written when he was in his third term. In it he had recounted the difficulties of the race. This made me wonder what made Jeetu run this grueling cross-country race again in the sixth term. What was it—Junoon or Fitoor?

These two Urdu words are fascinating in their overlaps. It is difficult to separate them in their underlying overtones. Fitoor and Junoon can be both obsession(s) with things, people, activity... I would like to believe that Jeetu had this Junoon to better himself every day as he was indefatigable. He was not there to prove anything to anyone; he was just pushing himself out of his comfort zone to improve continuously. I was peeling away hitherto unknown layers of my brother's personality long after he was gone. It was like reading a palimpsest.

Udhampur, being the headquarters of the Northern Command, is a hub of the army and there is constant movement of men and materials to and from the valley. Mom tells me how Jeetu's friends and batchmates would regularly descend on our Udhampur house and leave after having lunch or dinner. These were joyous moments for my mother too as the house would reverberate with their laughter and youthful energy. Today, too, some of his batchmates, if they are posted in Udhampur or are passing through, make it a point to visit our home. Umang Trehan, one of Jeetu's squadron mates, shared pictures of him meeting Mom. It is touching to know that Jeetu still remains in the memories of all those who entered his life during his brief journey on earth. And these friends with their memories keep visiting Mom.

The three-year training at the NDA ended in June 1998 with the award of a graduate degree from JNU. Mom, Vinkoo, Madhu, Aunt Kamla and I were present at the passing-out parade. At the grand parade, Shri George Fernandes, the then defence minister, was the chief guest. As usual, Mom with her elephantine memory remembers him entering the stadium clad in his trademark khadi kurta-pyjama and wearing chappals. Jeetu's graduation from the NDA was an emotional moment for us as we were missing Papa. It had been Papa's dream to see Jeetu as an army officer. He would have been so proud of his son.

The fellowships established in the NDA are strong but then paths diverge. Depending upon their choice—army, air force, or navy—the cadets go on to the respective service academies. Those opting for the army go to the IMA. Jeetu joined the IMA on 13 July 1998 and was in the Thimayya Battalion, Alamein Company. The Thimayya Battalion is named after General K.S. Thimayya, who served as the Chief of Army Staff from 1957 to 1961;

HAWS was also setup by him. Similarly, the Alamein Company is named after the Second World War battle of El Alamein in 1942, a crucial battle because it eliminated the Axis threat to Egypt.

At the IMA too Jeetu won medals for swimming as well as cross-country and his pen picture in their Spring Term 1999 magazine reads: 'True Mustang who deserved better tabs. Excellent in X country and swimming.'

Colonel Rajeev Bhola, Jeetu's roommate in the IMA, recalls, 'Shatrujeet was very focused and sometimes used to run alone to better his timing. In the first term at IMA, he came tenth out of 2,000 cadets and got a medal.'

Jeetu was commissioned in the Indian Army as an officer on 12 June 1999. All of us—Mom, Vinkoo, Madhu, Shivinder Mamu, Uncle Inderjit, Kailash Maasi and their young son, Varun—were present at the passing-out parade, which was held against the backdrop of the famous Chetwode House. The building is named after Field Marshal Philip Walhouse Chetwode, who established the IMA in 1932, one of the oldest military academies in India. Since the IMA's establishment, a total of 64,145 Indian Gentleman Cadets (GCs) and 2,813 foreign GCs from thirty-five friendly nations have been trained and commissioned as of December 2022 of the autumn term. I recall meeting Jeetu's batchmates from Bhutan at the ceremony.

Mom became emotional during the Pipping Ceremony. As she placed the lone stars on both shoulder flaps of

Jeetu's olive-green uniform, her eyes welled up with tears of happiness as well as sorrow. She was missing her husband and the father of her son at this momentous event in Jeetu's professional life. We sisters were so proud of our young, handsome brother. He was an officer in the Indian Army at a young age of twenty-two, standing straight and tall in his uniform as Mom pinned the stars.

After spending three years at the NDA and one year at the IMA, these young men, in their early twenties, are transformed into strong, decisive leaders dealing with matters of life and death. In very few professions has one had to shoulder such an onerous responsibility while so young. And I am so proud that Jeetu was one of their brightest. His name and photograph adorn the Hall of Fame in the Gallantry Awards page of the NDA's website. The text that accompanies the photograph reads:

MAJ Shatrujeet Kotwal
UNIT 3 JAT, COURSE-94, SQN-M
Year: 2008

In the photograph, Jeetu wears a solemn expression, viewing the world with equanimity. He was restrained when talking about his achievements as well as setbacks and took success and failure in his stride in equal measure.

Thus began his journey in the army, as an officer in the much-decorated and much-lauded, oldest infantry regiment of the Indian Army, the 3 Jat.

4

An Officer and a Gentleman

One who smiles rather than rages
Is always the stronger.

—Japanese proverb

In a short career of nine years (12 June 1999 to 17 June 2008), Jeetu achieved what many could not achieve in a lifetime. His decorations included the OP Vijay Medal, the Special Service Medal with the word 'Suraksha' (Sanskrit for safety or protection) inscribed on the clasp that keeps the ribbon and the medal together, the High Altitude Medal, the Sainya Seva Medal (with 'J&K' inscribed on the clasp), the GOC-in-chief commendation and finally the Kirti Chakra. All except the last two were news to me, and I came across these other decorations while going through Jeetu's papers. This is how he was, equanimous, humble, with his feet firmly on the ground and his head held high.

Jeetu had always wanted to join the army, and within

the army nothing but the infantry. My sister, in her letter dated 3 March 1995, writes, 'I want to tell you how happy I am about Jeetu's selection. My happiness knows no bounds. He had to be selected because he is a born fauji.' Our father was also an army officer but from the Electrical and Mechanical Engineering (EME) corps whereas for his son, it had to be the fighting arm, the foot soldier.

On graduating from the IMA in June 1999, Jeetu joined the 3 Jat Regiment. The Jat Regiment has been active since 1795, making it one of the longest-serving regiments in the Indian Army and also one of the most decorated. The motto of the Jat Regiment is 'Sangathan va Veerta' (Unity and Valour) and their war cry is '*Jat Balwan, Jai Bhagwan*' (the Jat is strong, victory to God). The Jats are known for their fondness for milk and ghee and are generally vegetarians. We used to tease Jeetu mercilessly on getting into the Jat Regiment, because he was a hardcore meat eater who loved his chicken biryani, mutton and karham (kohlrabi) in equal measure. A few days without meat, and he would develop withdrawal symptoms!

For Jeetu, joining the Jat Regiment was a matter of great pride and specifically the 3 Jat, which was raised in Dinapur on 23 June 1823. The regiment celebrated its bicentenary in September 2022, a year early, because they were moving to the northeast in October 2022.

The Battle of Dograi was the crowning glory of the

3 Jat, and Jeetu would get excited talking about it, particularly his meeting Brigadier D.E. Hayde (retd) in one of the regimental get-togethers. I had never seen my brother so excited about meeting another human being!

Brigadier Hayde, then a lieutenant colonel, was the CO in the Battle of Dograi. Nilesh Kumar, writing in the net edition of the *Indian Defence Review*, dated 23 September 2022, has recorded the reminiscences of Colonel Jha, who was part of the team and was a captain then.

The Battle of Dograi was amongst the bloodiest battles fought during the 1965 Indo-Pak conflict in which the 3 Jat created history by decisively defeating a numerically superior and well-entrenched enemy, and thus emerging as the highest decorated battalion of this war. Dograi is strategically important as it is just fifteen kilometres away from Lahore, the second largest city in Pakistan. The Pakistani army had exploited the locational advantage and the dense built-up area of Dograi to convert this medium-sized village on the eastern side of the Ichogil Canal into an unassailable defensive position.

This victory did not come easy as prior to their main battle on 20 September, the earlier attempt of the 3 Jat on 6 September had been foiled as their supply lines could not be reinforced because of strafing by the Pakistan Air Force. Between 6 and 20 September, the regiment made several attempts to capture Dograi that were unsuccessful as the village had been further reinforced with additional troops and interlaced with extensive minefields and wire obstacles. On 20 September, Lieutenant Colonel Hayde with 500 of his men was again tasked with capturing Dograi. Colonel Jha recounts: 'The night of September 21/22 saw brutal hand-to-hand fighting that continued nonstop for twenty-seven hours mostly in Dograi's narrow streets and dense built-up area. The sheer force of the 3 Jat's multi-pronged attack, with each of its four Rifle Companies simultaneously capturing the four corners of Dograi on the night of 21/22 September unnerved the enemy.'

The blood and sweat of the men of the 3 Jat created history but it came at a cost. In this seventeen-day long conflict, out of seventeen officers and 550 men that fought these battles, the regiment lost five officers, while nine (including Lieutenant Colonel Hayde) were wounded. One junior commissioned officer was killed while eight others were wounded; eighty-nine other ranks embraced martyrdom and 219 were wounded. This accounts for nearly 50 per cent of the total force that had joined this battle and gives an indication of the intensity of engagement. All this was not in vain, though: the capture of Dograi was just a day before ceasefire was announced and proved a valuable bargaining chip during the Tashkent negotiations that followed the 1965 war.

Colonel Jha has a special word of praise and deep respect for his commanding officer, Lieutenant Colonel Hayde, for having made this spectacular victory possible despite overwhelming odds through his innate tactical acumen, exceptional leadership qualities and most important, immaculate personal example.

I reckon all these qualities must have made Brigadier Hayde a legend, particularly among all the awestruck young officers who met him. After Jeetu's death, I made a list of officers to contact and in one of my diaries I had written down the landline number and address of Brigadier Hayde in Kotdwar. Unfortunately, I could never get in touch with him and he passed away in 2013.

My brother had kept in touch with him though. There is a neatly typed letter (except for the salutations written in hand) written by Brigadier Hayde to Jeetu on his letterhead dated 29 January 2008:

> Dear Kotwal, Delightful news! That of hearing of your award. I do recall the chats we had in passing when I visited the Bn at Ferozpur and we did indeed exchange letters, so this effort of yours renews the meaning of exchanges we had.
>
> ...I'm glad to hear that you are now an instructor at the NDA. You'll probably move from there to the staff college...

I am sure Jeetu must have been elated on receiving this letter from his hero.

With the Regiment in Uri

After leaving the IMA, Jeetu's first posting was in the Khemkaran sector, Punjab, where his regiment was stationed then. Colonel Aditya Verma, who would be Jeetu's colleague later on in the 34 RR during 2005-2007, recalls meeting him on and off during this period as he too was posted there: 'We were compatriots and would keep meeting on and off since 1999. In true military style, we would keep khabar of each other.'

Ranbir, Jeetu's orderly, tells me, '*Didi us time Kargil war bhi chal rahi thi aur situation tense thi* (The Kargil war was on and the situation was tense)'. He accompanied Jeetu from Khemkaran to Solan and finally to Uri in 2000. This was Jeetu's first tenure in the Kashmir valley. His regiment was stationed in Uri, a town known as the window of Kashmir because of its location. Before Partition, the Jhelum valley road linking Srinagar with Koh, Mari and Rawalpindi via Muzaffarabad entered the Kashmir valley from this town. At present, Uri is a subdivision of the Baramulla district of Jammu and Kashmir with two tehsils—Uri and Bunihar. Uri is about ten kilometres from the Line of Control (LoC) with Pakistan. Bakarwals, Pahadis, and Gujjars stay at the foothills, close to the LoC, amid the remote and far-flung areas of the valley.

Dheeraj was also posted to Uri then but in a different brigade. Both friends were on the same side of River Jhelum and close by. Dheeraj remembers that Jeetu had a terrific CO, Colonel S. Hasabnis, who came from a family of faujis. His father was in the army, and his younger brother, Lieutenant General S.S. Hasabnis rose to become the Deputy Chief of Army Staff, serving in that capacity from June 2018 to December 2020.

Colonel Hasabnis, who is now retired and lives in Pune, remembers Jeetu as 'a fantastic young officer who was physically extremely fit. He was disciplined, intelligent, and hard-working.' This is no mean compliment coming from a man who, according to Dheeraj, was amongst the most supremely fit senior officers and could challenge anyone on holding a self-loading rifle the longest.

'I had eleven young officers and seven were cross-attaches (those on secondment from other regiments). Shatrujeet was my adjutant and he did extremely well. He kept communication channels open with officers on one hand and the troops on the other hand. He had an open mind and I would get honest feedback from him,' recalls Colonel Hasabnis. 'He was a good fit for me as an adjutant. We interacted on a daily basis handling live situations and at the click of a finger, I would assign him multiple tasks. I could never doubt his capabilities to complete the assigned tasks.'

Very candidly, the Colonel quips, 'Most people know someone or the other about whom they don't like to talk about? But Shatrujeet was not like most people. I really appreciated the way he handled himself and his professional career. He was not the one to shy away from giving his honest opinion. I hold such youngsters in high esteem.'

What Dheeraj remembers fondly about Colonel Hasabnis is his love for milk and jalebis. On being reminded of it, Colonel Hasabnis laughs and says, 'I do not drink or smoke. So what better way to greet guests?'

'I used to call him Shatru and was ticked off by your mother when we had met at Jeetu's residence in Pune. She had corrected me, saying, "Colonel Sahib, his name is Shatrujeet and not Shatru. Shatru means enemy, and he is not your enemy",' adds Colonel Hasabnis with a chuckle. And that is our mother all over, never too shy to voice her opinion.

However, Colonel Hasabnis grows pensive, remembering the day Jeetu died. He was in Pune on the fateful day. 'He was an officer who would have commanded the regiment in his time but God had different plans.'

'*Didi bahut galt ho gaya, Sahib chala gaya. Woh tho regiment command karna tha* (a great wrong has happened that Sahib is gone. He was meant to command the regiment),' says Ranbir Singh woefully.

Dheeraj recalls, 'Post the Kargil War of 1999, the situation was tense around the LoC. Hemant Bisht was the first batchmate that we lost and then Gurjinder Singh Suri, our senior who was in the neighbouring sector. We were constantly on alert as there were many casualties being reported.'

The same thoughts were echoed by Brigadier Bodhe, who was a colonel then and had succeeded Colonel Hasabnis as the CO of the 3 Jat in Uri: 'Circumstances were quite grim at the LoC and we would be facing enemy fire from the Pakistan side almost daily.' He goes on to add: 'I have had thirty-six years of service in the military. However, I think being a company commander at the

LoC is the toughest task I have handled. Shatrujeet had a daredevil attitude but was not reckless. He was an excellent soldier, who planned ahead and understood his troops well.' Coincidentally, Brigadier Bodhe was also posted in Pune in 2008 and remembers with regret getting the news about Jeetu's passing away: 'On receiving the news, we got in action immediately and were there to complete all the formalities.'

Later on, the regiment moved to Ferozpur, still under the command of Colonel Bodhe, and Jeetu was his adjutant there. He was in Ferozpur till June 2005.

Like Colonel Hasabnis, Brigadier Bodhe also moved to Pune after his retirement and runs the Brig. Bodhe's Career Institute (BBCI), which trains young women and men for various competitive examinations conducted by the Union Public Service Commission (UPSC), for a career in the various branches of the defence forces.

During March 2000 and March 2003, while in Uri, Jeetu completed his YOs course at the Infantry School, Mhow, Madhya Pradesh; the commando course at the Infantry School, Belgaum; and basic and advanced courses in mountain and winter warfare. Though Jeetu did well in his YOs course conducted during March-July 2000 at Mhow, he missed his instructor grade by a whisker. After his course, he stopped in Delhi on his way back and I could sense that he was deeply disappointed by this, though he had completed the course with high grades. He suffered another setback when he missed his instructor grade again at the commando course during October-November 2000 at the Infantry School, Belgaum, by a narrow margin. He had done extremely well this time too, scoring a hundred per cent in physical fitness and endurance and over eighty per cent in commando technique and firing. However, the instructor grade proved elusive.

During my conversation with Dheeraj, I gathered that Jeetu and he were roomies at the YOs course. They shared a room for almost five months and also invested in a small scooter, which was christened 'Dhanno', a tribute to Dhanno of *Sholay* fame. 'She' proved equally faithful, ferrying the duo around the campus and beyond. They never said no to a brother officer, and sometimes a foursome would be riding Dhanno.

I told Dheeraj about Jeetu's reaction on his performance in the YOs and the commando course and his deep anguish. It was unlike him to be so deeply affected by setbacks. Now, speaking to Dheeraj, I could gauge the reasons for his disappointment at not making the instructor grade. 'For almost five months we put in four-to-five hours of study. I had never imagined that joining the army would make me study so much. There was an NCO, a havildar from the 3 Jat, who was posted to Mhow. He worked hard with us in improving our techniques. So, I feel Shatrujeet's disappointment also stemmed from the fact that he felt he had let down the havildar sahib from his regiment in not getting the instructor grade,' said Dheeraj. Similarly, during the commando course where Colonel

Hasabnis was also an instructor, Jeetu missed completing his rock-climbing test by two seconds. This test is less about strength and more about technique and has to be completed in sixty seconds. Today I can only conjecture that Jeetu was disappointed on two accounts, one on not getting the instructor grade and second, on not having performed too well before his ex-CO.

These two misses stayed with him.

There are two occasions that I remember vividly when he called me with palpable joy.

The first was sometime in September 2001, barely a year after his commando course. Jeetu had gone for the basic course in mountain warfare from July to September 2001 and was adjudged the best student among the officers. I guess this achievement was doubly sweet because it was a validation and vindication of his talent and hard work, and also a comeback after the setbacks at YOs and the commando course. It speaks volumes for the man, his steely determination, both physical and mental, his burning urge to excel and the ability to bounce back from adversity. He was recommended to attend the advanced course in mountain warfare, and the pen picture from the course report reads as follows:

> Capt. S. Kotwal is a sincere, hardworking and enthusiastic officer who took keen interest and achieved an overall High Average standard on the course. He displayed selfless devotion to duty and commitment to service interests of a very high order. He can effectively

> lead a team operating in Mountainous, High Altitude and Glaciated Areas. He was adjudged Best Student among officers.

He did extremely well in the Winter Warfare Basic as well as advanced courses that were held back-to-back from January 2003 to March 2003. His individual course report of the Winter Warfare Advance Course No. 44 has the following pen picture:

> Maj. S. Kotwal is an intelligent, enthusiastic and very hard-working officer, who achieved an overall Above Average standard on the course. He is physically tough, mentally robust and withstood in an excellent manner, the rigours of operating in high altitude, mountainous and glaciated areas. He is meticulous in his planning and can advise/and lead a team in high altitude, mountainous and glaciated areas. He has achieved proficiency in Skiing and Mountaineering and is a qualified instructor in these disciplines.

He was recommended to be an instructor at Category 'A' Training establishment/HAWS.

There is also a congratulatory letter dated 7 May 2003 from Major General S.S. Dhillon, VSM, General Officer Commanding, 19 Infantry Division, on Jeetu's outstanding performance: 'My congratulations at your outstanding performance during the WW Adv-44 course. The result is a reflection of your hard work and dedication. I am sanguine that you will continue to strive for further excellence in future. Well done and keep it up.'

Is everyone in the army driven like him? This was my question to Dheeraj. He replied candidly: 'From my experience, around eighty per cent work hard and twenty per cent are driven to excel. Shatrujeet was not afraid of getting his hands dirty at the basic grassroots level and was driven throughout his career beginning from training at the NDA.'

The second occasion when he called me with palpable joy was in June 2007, when he, in his circumspect manner, informed me that in an encounter they had killed a dreaded terrorist. On the phone, I could hear boisterous celebrations in the background and could sense the unbridled happiness and pride in his voice. Next day, this encounter was reported in the national dailies as a minor story and covered by DD news, the national broadcaster. But, more about this later…

Rashtriya Rifles: 2005 to 2007

Lieutenant General K.J.S. 'Tiny' Dhillon (retd), in his book *Kitne Ghazi Aaye, Kitne Ghazi Gaye: My Life Story*, likens the RR to a kimberlite stone that has metamorphosed into a diamond in an environment that demands the highest order of military precision and alertness from its soldiers 24/7 and 365 days a year. The Rashtriya Rifles is a counterinsurgency force raised for operations in Punjab during the 1980s. However, during the 1990s the RR became the main counterinsurgency force in Kashmir. According to Lieutenant General Dhillon, the RR has grown from its modest beginnings as simply being called a 'riff-raff force' by the naysayers to probably the world's most professional counterterrorism force. Rashtriya Rifles lives up to its epithet 'RR—*sirf naam hi kaafi hai*'—just the name is enough—as the most dreaded force by terrorists. The RR's crest, consisting of the Ashoka Chakra with two crossed bayonet-fixed rifles nestling against each other and overlaying its motto 'Dridhta aur Veerta' (Courage and Valour), is an authentic depiction of the concept of brothers-in-arms in the service of the nation.

Currently, RR has 65 battalions, and these battalions are affiliated to different regiments. For example, 4 RR is affiliated to the Bihar Regiment, 10 RR to Rajput, 13 RR to Kumaon, 21 to Guards, etc. Similarly, the 34 RR is affiliated to the Jat regiment and was then headquartered at Beerwah, a block in the Budgam district in Jammu and Kashmir. Budgam is bound by Baramulla district to the north-west, Srinagar district to the north-east and Pulwama District to the south-east. The Pir Panjal range separates Budgam from Poonch district to the south-west.

The officers and troops posted in the RR operate in difficult terrain. Some areas are densely forested, high mountainous regions with rugged stretches. Officers and troops are sent on deputation for about two or two and a half years to the RR. Jeetu was posted to the 34 RR in 2005 and remained there till 2007. It was his toughest as well as most rewarding tenure. The army has to undertake dual roles of guarding the LoC as well as countering terrorism. Jeetu played both the roles, first during his posting in Uri and then during his tenure with the RR. He was the company commander of the 120-strong Pethkoot Company.

Jeetu and Lieutenant Colonel Rohan, who was a captain then, joined the 34 RR around the same time in June 2005. Lieutenant Colonel Rohan, who is posted in Bathinda now, recalls: 'Shatrujeet Sir was senior to me and was the company commander. However, there was paucity of accommodation and we shared a room for almost two months. We had a good time and learnt about the RR together. He was extremely hardworking and motivated.'

With a chuckle that carries across the crackling phone line, Lieutenant Colonel Rohan reminisces: 'Shatrujeet Sir had an innovative approach to problem-solving and was a risk-taker too. He would tell me "Let us go for a walk". I remember Sir tucking his revolver under the tracksuit and both of us sauntering off. The area that we were operating in, namely the Kanadi belt, was quite notorious but Sir would be cool as a cucumber crossing the market and then stopping for chai and gup-shup at the village. *Bahut baar, hum log raat ko nikal jate* (many times, we would go out at night). Shatrujeet Sir was fearless.

'Shatrujeet Sir's batchmate Dheeraj Sir was also posted in the adjoining area and we would go across to meet him dressed in civilian clothes wearing pherans. Unlike us who were sharing rooms, Dheeraj Sir had a well-established set-up and would serve us good food. I remember both of them enthusiastically discussing their ideas on how to conduct operations over piping hot rajma-chawal,' recalls Rohan.

'Yes, we used to discuss various strategic issues. In CI operations people generally use aliases and not their real names. Shatrujeet and I did not change ours and we were known as the Kotwals. So, I guess people who were monitoring communication used to get confused, two different formations, the same person, who is it?' chuckles Dheeraj.

Lieutenant Colonel Rohan also distinctly remembers the earthquake that struck Kashmir in October 2005: 'It was morning, a little chilly and we were sitting in our room eating Maggi. Suddenly Shatrujeet Sir shouted, *bhookamp aa raha hai* (earthquake is coming). We rushed out and the ground seemed to move in waves. It was quite terrifying.' The devastating earthquake with a magnitude of 7.6 on the Richter scale struck at 8.50 a.m. on 8 October 2005. Its epicenter was a little over 19 kilometres northeast of Muzaffarabad in Pakistan-Occupied Kashmir (PoK). There was extensive loss of property and human lives on both sides of the border but the losses on the Indian side were comparatively low.

Around this time, both of them, Jeetu as well as Lieutenant Colonel Rohan, were sent to different places. 'Our CO then could not tolerate our bonhomie and asked why do we need two officers here?' remembers Lieutenant Colonel Rohan with a chuckle. He was sent to Toshe

Maidan, a high-altitude post, and Jeetu was moved to Tangdhaar. The Indian Army played a crucial role in rescue, relief and rehabilitation in the aftermath of the earthquake. The Tangdhaar Bridge had collapsed and as part of the rescue and relief mission, Jeetu with his troops had taken one month's supplies of essential items to the far-flung places in their area of operation.

Lieutenant Colonel Rohan also remembers with admiration Jeetu's capability in handling troops: 'Around sixty per cent of troops in the 34 RR were from various Jat regiments while the remaining were from artillery and other supporting corps. It is quite a task handling Jat troops. Jat troops are brave and courageous but are rough around the edges and will tell you straight to the face, *aisa tho nahi ho sakta,* Sirji (it cannot happen this way, Sir). There were officers from other corps who had problems handling the Jat troops but not Shatrujeet Sir. I think, him being from the Jat regiment helped and also, he was deeply empathetic and would advise me, "Understand human psychology, be patient, listen to them and try to read the minds of your troops."

'He would go to the langar, chat with the troops over tea or meals and mingle with an ease that came naturally to him. There was no artifice there. Even on some occasions when he got angry, he would maintain his cool and simply shrug and say, "chuck it". It was a learning experience watching him deal with the men. If he would find a jawan looking agitated, he would say, "*jawan pareshaan lag raha hai*" (the soldier is looking worried) and go across to him. He would speak calmly and quietly with him and try to understand what was bothering him, some issue at home or some family matter. Not many officers have that kind of patience,' remembers Rohan.

I also recall discussing with Jeetu human management and the incidences of fratricides that one would read about in the defence and paramilitary forces. Jeetu's astute observation then was, 'Mobile phones have added to the stress. Today a soldier is just a phone call away from being drawn into family feuds over land, property or any other personal matter. Earlier letters were our only means of communication and phone calls were a weekly phenomenon. Now, the stress transmission happens faster and so is the reaction of the person sitting hundreds of miles away from home at some remote border-post. He cannot do anything and feels helpless, frustrated. You need to feel the pulse of your troops, be calm while handling them so that they can trust you enough to share what is bothering them.'

Colonel Aditya Verma also shared the same thoughts: 'Shatrujeet was exceptionally cool and composed while dealing with the troops and was well-liked by his troops. They would say in clichéd terms, *Bada* royal officer *hai* (he is a royal officer).'

Ranbir tells me: '*Sabhi Sahib se itne khush the. Troops baat karte the Shatrujeet Sahib ke baare mein aur bolte, Sahib tho fauj ke layaak hai hi nahi. Kabhi garam hi nahi hota* (Everyone was so happy with Sahib. Troops would discuss and say Shatrujeet Sahib is not fit for the army as he never gets angry).' I guess that is the ultimate compliment coming from Jat troops.

Lieutenant Colonel Rohan observes: 'Beneath the calm, quiet personality and smile was a man with steely determination. Working with him was full-on training for me. In December 2005, I had done an operation and he was so happy. He hugged me and with a twinkle in his eyes told me, "I am sure, what I taught you has helped!"

'Another personality trait I remember was his openness to learn. He would share notes with seniors as well as juniors and was focused on his work. He would brook no digressions while on the job.'

Yes, Jeetu was a continuous learner. In his scribbling pad, he has noted:

Shape your present for your future...
Compete against yourself...
Believe IN URSELF.

It is sometimes so easy to sink into the bog of mundane daily living that we need those branches to pull ourselves up. We also keep reminding ourselves, the power of repetition. This is why I think worldwide self-help books are the fastest growing category of non-fiction books, and this trend has increased post the COVID-19 pandemic.

Colonel Bhatt of the 12 Kumaon Regiment was 2IC (second-in-command) of the 34 RR during 2005/06 and has a few memorable events to share. 'As the operation commander, we did few operations. I distinctly remember that on 26 May 2006, Aditya Verma, one of our officers, was engaged in an operation involving two terrorists and there was heavy firing. Quick Response Teams (QRTs) were deployed to support him. The company commander reached first, followed by the 2IC QRT. Shatrujeet's company, Pethkoot, was around 7-10 kilometres away from the area where the operation was taking place. On hearing the internal communication, he took upon himself to reach the place with his QRT. Here is an officer full of "josh" (fervour) ready to help his comrade and a true example of selfless service.

'On another occasion as officiating CO, I was leading a search party. It was a cold winter night. As we waited for further information from our sources outside a village, the troops built a bonfire. We all gathered around it, warming our hands. Shatrujeet was standing at a distance with his pack 08 weighing around 15-20 kilos on his back. I asked him to come near the fire. He neither came near the fire nor removed his pack. He simply smiled and said, "I am fine, Sir. Do not worry" and remained standing throughout the night, alert.'

Colonel Bhatt says with pride in his voice: 'I consider myself to be physically fit. I came first in the academy in the Battle Physical Efficiency Test (BPET) and ran 5 kilometres in eighteen minutes, thirty seconds. But Shatrujeet was equally good and very strong.

'Shatrujeet was brave, full of initiative, and ready to volunteer. He epitomized a perfect Indian Army officer. He was liked by everyone. We lost an excellent officer and a genuine human being,' says Colonel Bhatt wistfully.

Colonel Harjinder Singh joined as the CO of the 34 RR in January 2006 and he recalls: 'When I joined the 34 RR, Shatrujeet had already been there for a few months and he was the Pethkoot Company commander. We connected immediately. He was a sober, very mature and grounded officer. We were operating in a counter terrorism (CT) environment and he told me, "Sir, stability *laani hai* (we have to bring stability). If there is a terrorist group moving, we have to neutralize it."

'His company was in the interior and there was no tarmacked road. To break barriers with my officers, I would often visit him and he would tell me, "*Aap bhi bindass ho, Sir*" (You are also cool, Sir). It is extremely difficult and stressful working in a CT environment. Shatrujeet was a risk-taker but meticulous too. He had the vision, was thorough with his operational area and planned well. On one operation, he was out the whole night and I was on tenterhooks with the radio set on.'

Lieutenant Colonel Dass explains: 'To make one operation successful, one has to go through the rigour ninety-nine times, night and day, day in and day out. It gets very cold at night in the Kashmir valley and in the area of our operation, the temperature would drop up to −10 degrees. We would leave at 11 p.m. and be out

The Kotwal siblings (the author to the right, Vinkoo on the left and young Jeetu in the middle) with Suryakant Bhaiya, their father's orderly in Guwahati.

The Kotwals at Kotli Pain. *Top* The entire family, (clockwise from top left) Mom, Papa, Vinkoo, Jeetu, and the author. *Bottom left* Jeetu with Arjun. *Bottom right* All three siblings together.

Young Jeetu (7-8 years old) performing at a school function.

Winning the shield for inter-school debate competition.

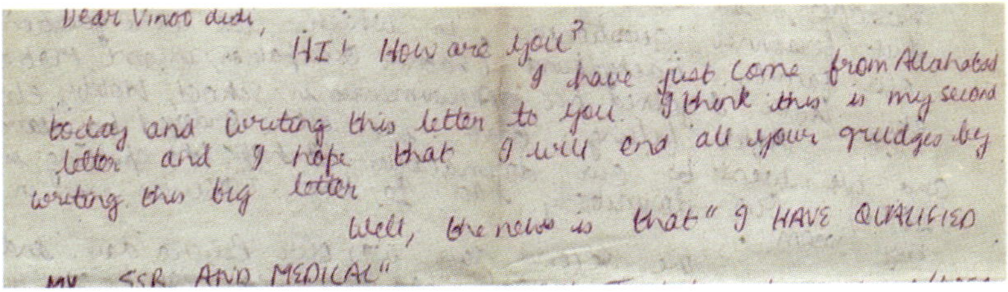

Dear Vinod didi,

HI! How are you?

I have just come from Allahabad today and writing this letter to you. I think this is my second letter and I hope that I will end all your grudges by writing this big letter

Well, the news is that "I HAVE QUALIFIED MY SSB AND MEDICAL"

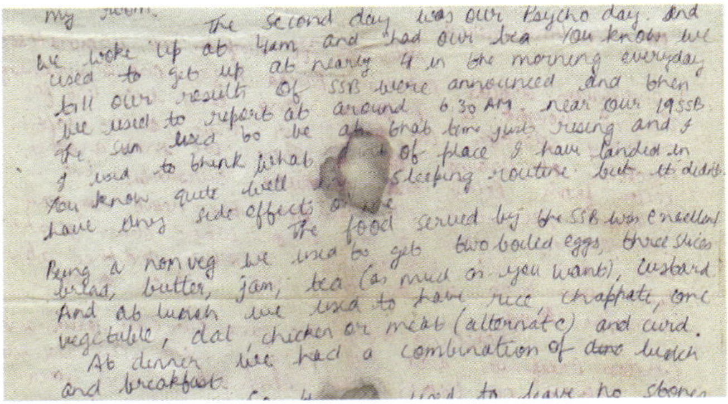

my room. The second day was our Psycho day. And we woke up at 4am and had our tea. You know we used to get up at nearly 4 in the morning everyday till our results of SSB were announced and then we used to report at around 6.30 AM near our 19 SSB. The sun used to be at that time just rising and I I used to think what [illegible] of place I have landed in. You know quite well [illegible] sleeping routine but it didn't have any side effects o[illegible] me.

The food served by the SSB was excellent. Being a non veg we used to get two boiled eggs, three slices bread, butter, jam, tea (as much as you want), custard. And at lunch we used to have rice, chappati, one vegetable, dal chicken or meat (alternate) and curd. At dinner we had a combination of lunch and breakfast.

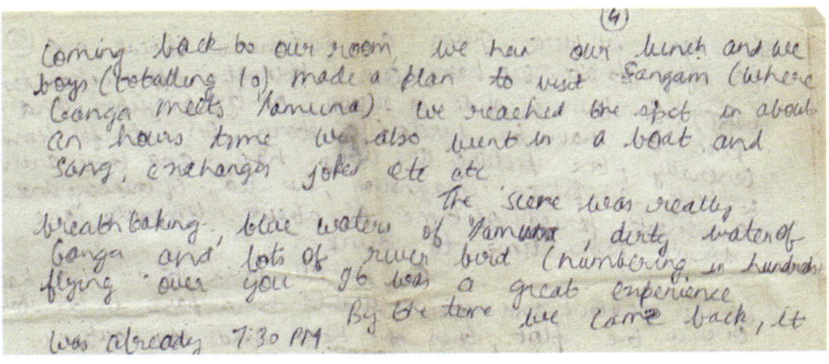

(4)

Coming back to our room, we had our lunch and we boys (totalling 10) made a plan to visit Sangam (where Ganga meets Yamuna). We reached the spot in about an hours time. We also went in a boat and sang, exchanged jokes etc etc.

The scene was really breathtaking, blue waters of Yamuna, dirty water of Ganga and lots of river bird (numbering in hundreds) flying over you. It was a great experience.

By the time we came back, it was already 7:30 PM

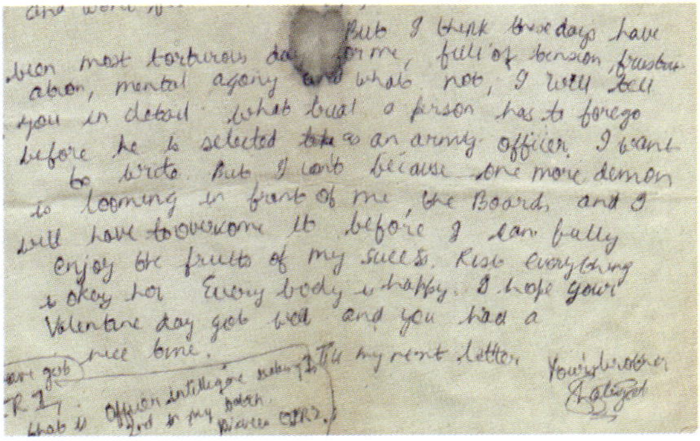

But I think these days have been most torturous da[illegible] for me, full of tension, frustration, mental agony and what not. I will tell you in detail what all a person has to forego before he is selected as an army officer. I want to write. But I can't because one more demon is looming in front of me, the Board, and I will have to overcome it before I can fully enjoy the fruits of my success. Rest everything is okay here. Everybody is happy. I hope your Valentine day got well and you had a nice time. Till my next letter.

Your's brother

Jeetu

A letter written to the author of Jeetu's experience at the SSB, Allahabad. Referred to extensively in Chapter 2.

Drill competition at the NDA. Jeetu, third from left in the front row.

'op The 'Mighty Mustangs' of Mike Squadron. *ight* Performing at the prestigious bayonet-ghting competition at the NDA.

Left Winners of a swimming competition at the IMA, Dehradun. *Below* Winner of the cross country medal at the NDA.

At the IMA graduation after the Pipping Ceremony. A proud mother with Jeetu and Vanu on right, Madhu on left, and Mamu and Inderjit Uncle in the background.

Jeetu the traveller. *Above* At the Kailash Yatra with Ranjeet and Venkat. *Below* In tradition gear in Rajasthan.

Forging new relationships: The Kotwal clan at Jeetu and Abha's engagement.

At the Line of Control near Uri. The Bravehearts of the 3 Jat.

A unit picnic of the 34 RR at Doodhpathri sometime in July-August 2006. *From left to right* Col Harjinder Singh, Maj. Shatrujeet Kotwal, Capt. Alok Dass, Capt. Choubey, Maj. Amit Pandey (bespectacled with glass in hand), Capt. Rohan (bearded with black sunglasses), Mrs Singh and Mrs Pandey. *Remaining three above* Maj. K.P. Vinay (with a glass on his head), Maj. Aditya Verma (bearded with a red cap) and Capt. Prabhat (with a grey cap).
Photo Courtesy: Major Aditya Verma

With his colleague from the 34 RR
at Beerwah, Srinagar.

At HAWS in Gulmarg for a training course. Jeetu was awarded the best student.

A proud hero. Smt. Pratibha Devisingh Patil, then President of India, pinning the Kirti Chakra on Major Shatrujeet Kotwal at the Investiture Ceremony on 7 May 2008.

DELHI

THE HINDU • SATURDAY, JANUARY 26, 2008

Ashok Chakra for Major Raghu Raman

He laid down his life fighting terrorists in Jammu and Kashmir

Special Correspondent

NEW DELHI: Major Dinesh Raghu Raman has been posthumously awarded the Ashok Chakra, the peacetime equivalent of the Param Vir Chakra, the country's highest award for gallantry.

President Pratibha Patil will confer the award at the Republic Day celebrations on Saturday. Three others — Capt. R. Harshan of 2 Parachute Regiment, Special Forces; Naib Subedar Chunni Lal, VrC, SM of 8 J&K Light Infantry; and Col. Vasanth Venugopal of 9 Maratha Light Infantry (all posthumous), who were awarded the Ashok Chakra on the eve of Independence Day last year — will also be honoured.

Born in 1978, Major Raman did his schooling in the Kendriya Vidyalaya, INA Colony, New Delhi, and joined the National Defence Academy in 1996. After passing out from the NDA, he was commissioned into 19 battalion of the Jat Regiment. He was awarded the Chief of Army Staff Commendation Card in Drass during the Kargil conflict.

Posted to 34 Rashtriya Rifles Battalion, he conducted a successful operation "OP Narawar" in June last year, leading to the elimination of three terrorists. On October 2, Major Raman deployed his company in a village in Baramulla district of Jammu and Kashmir, where an encounter with terrorists had begun. While closing in on suspected houses, Major Raman heard shouts of a fellow officer (Major K.P. Vinay, awarded the Kirti Chakra posthumously) who had been seriously injured.

He crawled towards the injured officer under heavy fire and shifted him and two others to safety. He then took on two terrorists who had caused heavy casualties to the troops and shot both of them dead. The other terrorists fired at Major Raman from another house, causing him injuries that proved fatal.

"Major Raman displayed the most conspicuous bravery, besides camaraderie and leadership of the highest order and made the supreme sacrifice for the nation," said an Army news release.

In all, the President has approved awards of 425 gallantry and other defence decorations to Armed Forces personnel and others.

Besides the Ashok Chakra, these include four Kirti Chakras, 22 Shaurya Chakras, two Bar to Sena Medals, 107 Sena Medals, two Nao Sena Medals, three Vayu Sena Medals, 30 Param Vishisht Seva Medals, three Bar to Ati Vishisht Seva Medals, 52 Ati Vishisht Seva Medals, one Yudh Seva Medal, three Bar to Vishisht Seva Medals, 127 Vishisht Seva Medals, four Bar to Sena Medals (Devotion to Duty), 38 Sena Medals (Devotion to Duty), 10 Nao Sena Medals (Devotion to Duty) and 16 Vayu Sena Medals (Devotion to Duty). One Tatrakshak Medal (Gallantry), two Takrakshak Medals (Meritorious Service) and 28 Mention-In-Despatches have also been announced.

Major Raman

Major K.P. Vinay

Major Shatrujeet Kotwal

Kirti Chakra

Among the Kirti Chakra awardees, two have been given posthumously to Major Kumandur Prabhakar Vinay of 53 Medium Regiment (Artillery) and Sepoy Suresh of 16 Jat Regiment.

Major Vinay from Hyderabad was an alumnus of the National Defence Academy. He killed two terrorists in Baramulla district on October 2 last year before succumbing to his injuries. He is survived by his father K.R. Prabhakar.

Sepoy Suresh of 16 Jat Regiment was martyred in an action with militants in Kupwara district of J&K on October 6 last year. He killed three militants even after being injured in a firefight with them.

Of the two surviving Kirti Chakra winners, Major Shatrujeet Kotwal, while serving with 34 Rashtriya Rifles, also killed three terrorists in a daredevil action in Budgam district (J&K) in June last year. An alumnus of the NDA, he was commissioned in 1999. He belongs to Udhampur (J&K).

The second surviving awardee, Major Punia led the assault against militants in Manipur who were entrenched in a camp. He killed two militants and along with his colleague killed another and injured one more.

Above Major Shatrujeet Kotwal was awarded the Kirti Chakra with two of his colleagues from the 34 RR, Major Raghu Raman (AC) and Major K.P. Vinay (KC). *Below* Major Shatrujeet Kotwal standing tall while the citation is being read at the Investiture Ceremony.

Ø1382-222136

IX

JAT REGIMENT

THIRD BATTALION

Brigadier D E Hayde
Bharati Bhawan,
Kotdwara-246149

29 Jan '08

Dear Kotwal

Delightful news! That of hearing of your award. I do recall the chats we had in passing when I visited the Bn at F'Pore and we did in-deed exchange letters, so this effort of yours renews the meaning of the exchanges we had.

Your charming wife must be giving you sops for sticking your neck out, now you can let her understand how much you will receive in return morally, financially and in a few other factors.

I'm glad to hear that you are now an instructor at the NDA. You'll probably move on from there to the Staff College, for which reason I'm sending you a copy of a recent script I drew up, as a last gasp one might say, since it has irked me that what 3 Jat did on 6 Sep '65, when all the rest of our side crumbled, has not been appropriately recognised.

Commandants Army War College and DSSC have both confirmed that the material herein is being inculcated in their training syllabus. In any case, while at the NDA and being a 3 Jat Officer you may have to interact with others during Mil Hist schedules.

Regards to you both
& again
congratulations
As Ever
Desmond Hayde.

Major Kotwal, KC
NDA Khadakvasla

A heartfelt letter of congratulations from one of Jeetu's heroes, Brigadier Desmond Hayde.

Jeetu with the two loves of his life:
Above with Abha.
Left with Shilpa.

Major Shatrujeet Kotwal, KC in military regalia, looking charming and smiling luminously at his Pune home.

the whole night not once but many times. A sense of frustration would creep in on not being successful and Shatrujeet would say, "*Ho jayega, tension kyo leh raha hai?*" (It will happen, why are you getting tense?) In the middle of an operation under extreme pressure, he would be calm like a still lake. He would absorb the pressure, coordinate everything and then in his laconic manner say, "Let us finish and go and have a beer". He was a step ahead of us.'

I also recall Jeetu recounting the experience of manning those posts deep in the mountains surrounded by dense forests in winter, those night reconnaissance missions with his men, especially of his tenure in Uri. It would be freezing and they would be walking quietly but always vigilant, with their backpacks and guns slung around the shoulders. The fir or the pine needles on the forest floor would muffle the footsteps but any rustling branches would put them on high alert. He was barely twenty-three when he was leading his men in those life-and-death operations, not knowing what was lurking around the next gully or ridge. These young men are trained well. However, those are simulated conditions, and experiencing and reacting in real-life situations on a daily basis can be extremely stressful.

Lieutenant Colonel Dass goes on to add: 'I am a corporate trainer now and give lectures on leadership styles, how to manage stress. According to me a good leader has maturity and a commonsensical approach to solving issues without any grandstanding.

'Shatrujeet had both. I remember an incident in this context very clearly.

'We used to have a conference at the battalion HQ once a month and all the company commanders would be there. For security reasons, we would move before dawn and leave after lunch before it got dark. Generally, the company commander would have a vehicle. However, while returning back to the company from the battalion HQ, there would be more people as many troops would be returning from leave. Hence, there used to be a tussle between the company commander and the in-charge JCO of transport at the battalion HQ on requirement of an extra vehicle.

'A similar situation arose once when the Subedar JCO Sahib was not ready to provide an additional vehicle for Shatrujeet's company. After the battalion conference, all the company commanders were lounging in my room after lunch and Shatrujeet was informed that additional vehicles cannot be provided. On hearing this, without batting an eyelid, Shatrujeet told his men to disembark from the vehicle with their luggage. "It's been long since we did patrolling. Everyone will pick up their luggage and we will walk back together to the company." As his men started dismounting from the vehicle with their luggage, the Subedar JCO in-charge of transport came running to Shatrujeet, "Sir, we have been able to arrange vehicles and they are ready." Shatrujeet looked at him, smiled and said, "Thank you Subedar Sahib."

'We were in splits after Subedar Sahib left. Subedar Sahib could not afford to have one of the company commanders with his men trek seventeen or eighteen kilometres in the dark. One of the seniors looked at him and asked quizzically, "Kotwal, *tere dimaag mein yeh kaisa aaya?*" (Kotwal, how did this occur to you?) Shatrujeet smilingly replied, "Sir, *yeh sab common sense hai.*" (Sir, this is all common sense).

'I never thought I would use this incident in my corporate training many years later. This is how Shatrujeet was, a natural leader.'

These men living and working together under such stressful conditions also occasionally take out the time to relax. Colonel Aditya Verma shared a picture of the unit picnic at Doodhpathri sometime in July-August 2006. The name Doodhpathri means Valley of Milk. It is an alpine valley framed by snow-clad mountains and covered with meadows of pine, fir and deodar. In spring, the natural meadows, which are blanketed with snow in winter, would burst into colour with wild flowers such as daisies, forget-me-nots and buttercups. I can't see much of the beautiful meadows in the picture but what stood out were the smiles on everyone's face. Colonel Harjinder Singh, accompanied by his wife, is leading the smiling brigade of Jeetu, Captain Alok Dass, Captain Choubey, Major Amit Pandey and his wife, Major K.P. Vinay, Major Aditya Verma, Captain Prabhat and Captain Rohan. It is a beautiful picture radiating sunshine, warmth and capturing the strong bonds between these men. And I was fortunate enough to speak to four of them on this journey of remembrance.

Colonel Harjinder Singh who himself was remembered by his officers for being passionate and compassionate, recalls Jeetu's humane nature: 'In a CT environment, it is crucial to have ground-level information and thus, building bonds with the local communities becomes extremely important. If this link is broken, conducting operations becomes difficult. As a part of intelligence gathering, we have khabaris (informers) in the surrounding villages. Relationships with them have to be built and nurtured, as they provide us with valuable information on the movement of terrorists. Once one of Shatrujeet's informants got knocked off and he was very upset. He wanted us to help his family as much as we could. In ensuing months, he tried hard to find the killers. That loss stayed with him. Not many officers bother so much. But Shatrujeet was a compassionate man and cared not only for his men but for all those who worked with him.'

Listening to all these stories about Jeetu from his seniors and colleagues also made me realize that our brother had a high 'emotional intelligence' (EI). Daniel Goleman, an American psychologist and author, through his critically acclaimed books on EI argued that non-cognitive skills can matter as much as the Intelligence Quotient (IQ) for workplace success and leadership effectiveness. According to Goleman, 'Emotional intelligence is the *sine qua non* of leadership. Without it, a person can have the best training in the world, an incisive, analytical mind, and an endless supply of smart ideas, but he still won't make a great leader.'

So, what are the five components of EI at work?

Self-awareness, self-regulation, motivation, empathy and social skills.

Jeetu had them all.

In our short meetings over the years, I was unable to fathom the hidden depths in Jeetu's personality. As they say, still waters run deep. Here was a young man in his twenties, leading his men in life-and-death situations on a daily basis with dignity and courage. I am sure it would not have been easy. Balancing professional responsibilities with personal commitments as a son, husband, brother...

The image of these testosterone-fuelled men has another dimension. Dheeraj concurs: 'Being in the field is a huge responsibility and can be stressful. One is on duty 24/7, 365 days a year and sometimes one battles insomnia. You are continuously in a state of high alert, either you are in the field yourself or you are monitoring the movement of your men in the field. Being in a constant state of alertness leaves a lasting impact on you.'

These stories are not unique to Jeetu; they are of each one of them battling it out day and night at the LoC or in a CT environment. We idolize the unicorns; raise toasts to young entrepreneurs in chandelier-adorned ballrooms in bustling metropolises. I think it is high time we salute these brave men and women of our armed forces and at the same time, speak about their vulnerabilities too!

During Jeetu's tenure in the RR, he was sent on two additional courses, the All Arms Officers Combat Engineering course (21 March to 30 April 2005) in Pune and the Junior Command course (8 October to 1 December 2007) at the Army War College, Mhow. In the course in Pune, four participants out of a total of fifty-two secured grade 'A', and Jeetu was one of them. In his Junior Command course too, he earned the grade of Instructor for a category 'A' establishment.

While going through Jeetu's career trajectory, I was impressed by the deeply ingrained ethos of the Indian Army to train its men and women continuously, and felt that this is one thing worth emulating by other organizations. During his short career, Jeetu was trained rigorously and what stood out was the genuine feedback that was given, the nurturing attitude of the senior officers and the acknowledgement of their hard work. I am sure this must be the same in other branches of the defence forces too, thus making the Indian defence forces one of the best.

For Jeetu, the big news came on 25 January 2008 when the list of winners of the gallantry awards was announced on the eve of Republic Day, a recognition for killing a dreaded terrorist in an encounter in June 2007.

5

Guts and Glory

When can their glory fade?
O the wild charge they made!
All the world wondered.
Honour the charge they made!

—Lord Tennyson
'The Charge of the Light Brigade', 1854

On 25 January 2008, the Press Information Bureau, Government of India, on behalf of the Ministry of Defence, released the Republic Day list of gallantry and other defence decorations. The President Smt. Pratibha Devisingh Patil had approved awards of 425 gallantry and other defence decorations to armed personnel and others. These included one Ashoka Chakra (AC), four Kirti Chakras (KCs), twenty-two Shaurya Chakras (SCs) and 107 Sena Medals amongst others.

Officers from the 34 RR had been awarded three out of the five highest gallantry awards, one AC and two KCs.

Major Dinesh Raghu Raman of the Jat Regiment was awarded the Ashoka Chakra posthumously; the KCs were awarded to Jeetu and, posthumously, to Major Kumandur Prabhakar Vinay of the artillery regiment. Other KC awardees were Major Subhash Chand Punia of the 21 Parachute Regiment (Special Forces) and, posthumously, Sepoy Suresh 16 of the Jat Regiment. Men from the Jat Regiment were living up to the highest standards of valour.

The next day, the leading newspapers carried the news with varying headlines. The *Times of India*, Delhi edition, announced '4 get Kirti Chakra, 22 Shaurya Chakra today'; the *Indian Express*, New Delhi edition, proclaimed 'Major who died in Baramulla gunfight gets Ashok Chakra'. The *Hindu* carried a five-column-long story under the heading 'Ashok Chakra for Major Raghu Raman'. Along with Major Raman, the story featured photographs of two other awardees—Major K.P. Vinay and Major Shatrujeet Kotwal.

While recounting his tenure as CO of the 34 RR, Colonel Harjinder Singh was both proud and sad in equal measure—proud of the achievements of his officers, sad to have lost three of them in less than a year.

Major Dinesh Raghu Raman and Major K.P. Vinay were killed in an operation on 2/3 October 2007; Jeetu died on 17 June 2008. The timings of their deaths hold a kind of universal irony—the best laid plans of mice and men, if you may. Major Raman was to proceed on leave to meet his wife in Delhi for the birth of their first child.

Major Vinay was also to go on annual leave in November as he was getting married. However, life has other plans for you, over which you have no control. Is there any wisdom living for a future over which one has no agency, or in the past which is gone? Thus, the philosophy of *carpe diem* appeals to me immensely.

Of the three friends from the 34 RR, only Jeetu received his medal 'in person'. The Defence Investiture Ceremony was held at Rashtrapati Bhavan on 7 May 2008. Abha, Mom, Abha's parents, Madhu and I were present at the majestic Ashok Hall to witness Jeetu receiving the Kirti Chakra. The ceremony was conducted in a solemn manner with dignity and grace, helmed by the Hon'ble President of India, the Supreme Commander of the Indian Defence Forces. As we stood up for the national anthem, I was overwhelmed with swirling emotions. Here was a boy from a small town with 'fire in his belly', standing before the Hon'ble President of India, Smt. Pratibha Devisingh Patil, ready to receive his gallantry medal. Our little brother, through his hard work, fortitude, dedication, devotion to duty and bravery, had reached the pinnacle of his career.

Before he received the medal from the president and while the citation was being read, Jeetu had stood ramrod straight with arms by his sides, fists clenched, with an inscrutable expression on his face. He was in his full army regalia with the ceremonial cap and four medals pinned proudly on his chest—the OP Vijay Medal, the

Special Service Medal, High Altitude Medal and Sainya Seva Medal. He had looked solemn and exceptionally handsome that day. I guess this is a sister's bias speaking but I'll have that image etched in my mind forever. Seated in the front row, right behind Jeetu was Dr Manmohan Singh, the then Prime Minister of India; Smt. Gursharan Kaur, his wife; Shri A.K. Antony, Defence Minister of India; and other dignitaries.

The investiture ceremony citation read as follows:

> *MAJOR SHATRUJEET KOTWAL (IC-58609)*
> *JAT REGIMENT/34 RASHTRIYA RIFLES*
>
> *On 18 June 2007, Major Shatrujeet Kotwal received information that a dreaded terrorist had escaped along with other heavily armed terrorists in general area in Budgam district of Jammu and Kashmir. The officer with his small team entered a deep Nala in pursuit. The terrorists retaliated with heavy indiscriminate fire. This officer, however, closed in and killed two terrorists in a daring act. However, a third terrorist tried to escape under the cover of heavy fire. Major Shatrujeet Kotwal putting his own life at grave risk spontaneously accosted him and shot him dead at close quarters.*
>
> *Major Shatrujeet Kotwal displayed a high degree of professional competence and conspicuous gallantry in fighting the terrorists.*

As a crowning glory, there is also a photograph of the President of India pinning the KC to Jeetu's chest. This photograph now decorates a wall of our Udhampur home as well as our home in Delhi.

Jeetu would be deeply distressed if I do not mention the Ashok Chakra citation honouring Major Dinesh Raghu Raman and the Kirti Chakra conferred on Major K.P. Vinay, his colleagues and friends in the 34 RR.

DINESH RAGHU RAMAN
19th BATTALION (THE JAT REGIMENT)
(POSTHUMOUS)

On 02 October 2007, at 0820 hrs Major Dinesh Raghu Raman deployed his company in a village in Baramulla district of Jammu and Kashmir where an encounter with terrorists had commenced. At 0855 hrs, while closing in on suspected houses, Major Raman heard shouts of a fellow officer who had been seriously injured. Displaying camaraderie and esprit-de-corps of the highest order he crawled towards him under heavy hostile fire. He extricated him and two other injured soldiers to safety. He then took on two terrorists who had caused heavy casualties to the troops and in a fierce close quarter battle shot both of them dead. The other terrorists now fired on Major Raman from another house. In the ensuing firefight, Major Raman was seriously injured. Undeterred, he continued leading and motivating his troops and pinned down the terrorists till he fell unconscious. He later succumbed to his injuries.

MAJOR KUMANDUR PRABHAKAR VINAY (IC-58793)
REGIMENT OF ARTILLERY/ 34 RASHTRIYA RIFLES
(POSTHUMOUS)

On 02 October, 2007 at 0830 hours Major K.P. Vinay volunteered to lead a Ghatak Platoon to a village in Baramulla district of Jammu and Kashmir where an encounter had commenced. A young girl and a lady had been taken hostage by the terrorists. Major Vinay, leading from the front, crawled towards the suspected house. As he came close to the house three terrorists rushed out firing at him. Major Vinay however stood his ground and in a close quarter battle killed two of them on the spot. During this exchange of fire, he was hit by a burst of bullets and made the supreme sacrifice.

Major K.P. Vinay displayed conspicuous bravery and exemplary leadership qualities in fighting the terrorists.

The fellow officer referred to in Major Raghu Raman's citation was none other than Major K.P. Vinay. These men were bound in life and in death.

As usual, it is my mother who remembers that Major K.P. Vinay's parents were present at the Investiture Ceremony, and that it was his father who received the medal on his behalf. While the KCs and SCs were presented on 7 May 2008, Major Raman's Ashok Chakra was presented to his father by the President, earlier on Republic Day, 26 January 2008.

With the roll of drums, the ceremony came to an end and the medal recipients were conducted to the Fountain Loggia for group photographs with the President, the Vice President, the Prime Minister, the Defence Minister and other dignitaries. This was followed by tea and light refreshments at the banquet hall and at Upper Loggia. The entire atmosphere was charged with emotion and energy. The ceremony, which was held in the evening started at 1830 hrs. By the time it ended, night had fallen.

We met Colonel Harjinder Singh, the proud CO of the 34 RR, at the ceremony. During his tenure in 2008, apart from receiving the AC and two KCs, five Sena medals were also awarded to Colonel Harjinder Singh, Major N.C. Gairola, Major Aditya Verma, Subedar Satish Kumar and Naik Desh Raj.

The details of the investiture ceremony were extensively covered by many newspapers. One of them noted:

> The late Major Prabhakar Vinay and three others—Major Shatrujeet Kotwal of the Jat regiment, Capt. Pankaj Kumar of the Gorkha rifles and Sepoy Suresh—were awarded Kirti Chakras, the nation's second highest peacetime gallantry award in the military roll of honours.
>
> Major Kotwal pursued an escaped dreaded terrorist and raided their hideout in the deep jungle in Budgam district, killing two terrorists...

These conventional lines do not capture the drama and the intensity of the whole operation. I had reached out to

Colonel Harjinder Singh when I started writing this book, after having received his mobile number from Dheeraj. And it was Colonel Harjinder who had further shared the contact details of other officers who were posted in the 34 RR at that time. In this way, a chain of remembrance was being woven.

Colonel Harjinder recalled with extreme clarity not only the details of the operation where Jeetu was involved but also the Baramulla operation involving Major Raghu Raman and Major K.P. Vinay.

'We had received intelligence reports that Lashkar-e-Toiba (LeT) was on the rise for the past few months and was trying to revive its cadre with an aim to carry out some disruptive acts in the Budgam district. The group was being traced by the Jammu and Kashmir police and our own sources for the past one month. Constant contact was being maintained by me with the Senior Superintendent of Police (SSP), Budgam, on the activities of the group.

'There was a political rally planned by Ms Mehbooba Mufti, the leader of the People's Democratic Party (PDP), in Beerwah on 18 June 2007. We received information that a senior commander of the LeT was planning to attack the rally. There were three access points to reach the venue and in close coordination with the Jammu and Kashmir police, we made a plan to trap the terrorists. Small teams were placed on road junctions such as Sonapath, Lalpura and between Aripanthan and Qasba Biru bus stand covering Sukhnag. Shatrujeet's QRT was moved from interior

Pethkoot; he was hiding himself but keeping an eye on the movement of men and vehicles. Raghu Raman's QRT was at Aripanthan, and the 2IC QRT was centrally placed at Beerwah. The CO's QRT looks different and we did not want to draw any attention, so that was placed on standby at the Battalion HQ. We did everything quietly and swiftly, making sure that things looked normal from the outside. I also left as if on routine movement. But, I was in touch with everyone.

'The rally was to take place in the afternoon. Around 1530 hrs, Shatrujeet reported that the Station House Officer (SHO) had crossed him with his people in great hurry in a vehicle. They were driving furiously towards the village Narawar. We were wondering, "Why are the police leaving the political rally? Maybe they have got some information as the police also have their own network of informers."

'Knowing the SHO well, we were sure that some definite information was there. Shatrujeet and his team followed the police party. I told Raghu Raman to leave for Narawar immediately too. In the meantime, around 1600 hrs, Shatrujeet informed that there was firing in Narawar. We got the information that two foreign terrorists and one local had left Narawar after firing.

'Shatrujeet and his team had moved in civil-hired transport (CHT) up to the village Mirpura and from there they entered the Mirpura-Narawar Nala on foot to cut off the escape routes. A team under Raghu Raman moved from another direction, that is, from the village Chiuyadara towards Narawar. Subedar Satish along with the commando QRT also left. I also moved with my QRT to Narawar. In the meantime, we lost contact with Shatrujeet's team.

'Around 1630 hrs, we were able to re-establish contact with Shatrujeet's team and they were at Narawar Nala. At places the nala is deep, almost like a gorge; between 1630–1645 hrs heavy firing broke out in the nala. Shatrujeet and his team kept following them. There was firing from both sides. Shatrujeet reported, "Sir I can see them running." Firing was not effective and we started lobbing grenades. In the meantime, Subedar Satish's QRT reached the centre of the nala and I too reached there in my vehicle. I followed the Shatrujeet's tail, who was chasing two terrorists, one of whom was injured. Raghu Raman was also closing in.

'Shatrujeet was physically fit, he followed the terrorists swiftly and knocked them out face to face. He and his team displayed cool nerves, tactical maturity and eliminated the terrorists in fifteen minutes flat. In the meantime, the local guy slipped into a smaller, deep nala. He was actively pursued and eliminated too.

'It was confirmed that one of the two foreign terrorists killed was the Divisional Commander, LeT. He was definitely planning to do something at the political rally. This was corroborated by the Intelligence Bureau (IB) reports too. We recovered one AK-47, two AK-56s, three

AK magazines, four Under Barrel Grenade Launchers (UBGLs) and sixty-four rounds of AK ammunition.

'I still remember vividly that after the incident, when Shatrujeet was climbing up from the nala, he looked totally charged. When our eyes met, we exchanged big smiles. As he came close, we shook hands and hugged. Had we been a little late in engaging, the terrorists would have escaped. After the incident, we met the Deputy Inspector General (DIG) at the police station, completed all the formalities and handed over the three bodies and the weapons to them.

'Next morning, I got a call from Lieutenant General H.S. Panag, GOC-in-C Northern Command. He said, "I am coming to Awantipora and would like to meet you and the team." Three days later, Shatrujeet, QRT NCO and two more boys, one from Satish's QRT, and I left to meet Lieutenant General Panag. Shatrujeet and I drove together. He was looking very smart and handsome that day. We met Lieutenant General Panag at the airfield and I introduced Shatrujeet as the officer who had led the operations. We then walked down to the conference hall for a briefing. I personally briefed Lieutenant General Panag in detail about the operation, particularly our intelligence-gathering efforts for the past few months.

'Appreciating the team effort, outstanding bravery and devotion to duty, Lieutenant General Panag awarded Major Shatrujeet Kotwal, Sepoy Ajay Kumar, Sepoy Naresh Kumar and Lance Naik Mangal Chand the GOC-in-C Northern Command commendation cards on the spot.'

In my earlier discussions with Lieutenant Colonel Rohan, he too recalled this meeting with GOC-in-C Northern Command as he was stationed at Awantipora then. He clealy remembered meeting Colonel Harjinder Singh and Shatrujeet at the airfield and remarked how dashingly handsome Jeetu was looking that day.

Years later, during our conversation on 4 August 2023, Colonel Biswajit Dash told me, 'During Shatrujeet's operation in the 34 RR, I was posted in the Northern Army HQ at Udhampur. His name had been recommended for the Shauraya Chakra for the operation. However, Lieutenant General Panag, the then GOC, was highly impressed by the way the whole operation had been conducted; its high degree of professionalism and precision. He upgraded and recommended Shatrujeet's name for the Kirti Chakra.'

Colonel Harjinder also remembers with clarity another occasion on 2 October 2007, a few months after the incident in Narawar.

'A group of nine foreign terrorists had come from Poonch via Lahore Mandi and had crossed the Pir Panjal. The terrorists were tired and hungry and wanted food and a place to rest. At Poonch they encountered a young local boy who used to tend mules and had done some work for us. They threatened the young boy and he escorted them to the village Rengawari, which was in our area of operation. After leaving the terrorists at Rengawari, the boy ran back and informed the first fauji he met about the terrorists. The fauji in turn reported it to the Subedar Sahib who immediately left with seven or eight soldiers for Rengawari. While nearing the village they were fired upon. Raghu Raman was about one hour away from the operation site.

'It was the morning of 2 October 2007. Major K.P. Vinay, who was the adjutant, broke the news of our unit's, the 34 RR (Jat), contact with these freshly infiltrated Pakistani terrorists. Vinay was directed to take stock at the Battalion Headquarters and get the QRTs mobilized. But, he insisted on leading one of the Ghataks, which he finally did. While Vinay was moving to the operational area, Raghu Raman was en route as per our plan. Both were about thirty minutes ahead of me on the way to the place of contact. When I reached the location, the first information received was that K.P. Vinay and Raghu had been shot while fighting and leading their men from front.

'All hell broke loose thereafter since the terrorists were not allowing us to retrieve Vinay's body. After great difficulty, we managed to get him out, but during the bargain, two more boys (Navneet and Harpreet) along with Raghu Raman were injured. K.P. was dead by the time I took him in my arms. However, Raghu had a bullet injury and was immediately evacuated to the hospital, where he died. While being taken to the hospital, he gave me valuable information and kept saying, *K.P. Sir ko mara hai, in ko chodna nahi* (They have killed K.P. Sir; do not let them go).

'The Jats were down, but furious. It took almost an hour to stop the firing. Another hour or two to readjust. But my team leaders, one who had joined from the BB Cantt (Srinagar) after hearing the news on TV, were sure and ready to settle the score.

'By now it was evening. The boys were mobilized to get kerosene lanterns (Petromax) from neighbouring villages and generators from the Battalion Headquarters. I, for sure, knew that the actual action would happen at night since the terrorists would like to sneak out under the cover of darkness. At midnight, three or four desperate attempts were made to break the cordon. However, the Jats played well, displaying valour, calmness and bravery. Reports of dead bodies started trickling in. I kept my fingers crossed hoping and praying none were civilians.

'Around 3 a.m., one of the terrorists broke our cordon. A cool and calm officer, who later commanded the 34 RR, followed him, risking his life under mortar illumination, and knocked him down in a maize field.

'At first light, there was total mayhem since every staff officer wanted to know the number of kills. Since I had not physically seen the bodies, I refused to commit to numbers. The General Officer Commanding was on the line and wanted figures but couldn't get any. By 7 a.m., the Army Commander was on the wireless set. He said he was mobilizing Para Special Forces. I reasoned with him that I had lost two officers and it was our responsibility to settle a score. I assured him that he would not be let down but he needed to give us time. He agreed and wished me well.

'By three in the afternoon of the next day (3 October), when I gave my first report that we had nine dead, all Pakistanis, no one believed me. Corps Cdr who had retired for the day asked for a chopper and was with us in no time. When I received him, he asked again if I was sure of my count. In the evening, the Army Cdr was on the line; he said, "I am proud of your unit, since you have lived up to your commitment; well done."

'Immediately after the operation, I got my officers together and asked about their recommendations for their comrades. In true spirit of the Olive Greens, they agreed that the highest respect be given to the fallen soldiers for valour, raw courage, going beyond the call of duty and laying their life in the service of the nation. Had they not rushed in and pinned them down, all nine terrorists would have gotten away.'

Colonel Harjinder Singh continues pensively: 'I lost three officers, who represented the finest of the Indian Army. They were "General" material. Half the things in the RR, I learnt from them. Rest in Peace.'

On the evening of 13 September 2023, there was a newsflash on the killing of three officers, two from the army and one from the Jammu and Kashmir police, in an operation in Anantnag district. Suddenly in my head, I could hear Colonel Harjinder recounting the two operations. After sixteen years, the events of October 2007 were repeating themselves in 2023 with different people in the same setting. A feeling of déjà vu washed over me.

Next day, on 14 September 2023, the electronic media was awash with details about the three of them—Colonel Manjeet Singh (41) and Major Aashish Dhonchak (34) from the 19 RR, and Deputy Superintendent Himanyun Muzamil Bhat (29) of the Jammu and Kashmir police. There was a passing mention of Rifleman Ravi too. Another human tragedy was unfolding before my eyes and I could foresee what lay ahead for the families of these men. It would take years for the families to come to terms with what had befallen them. Bhat, the youngest, was married the previous year and had a two-month-old daughter. Similarly, Colonel Manjeet Singh had two children, aged six and two years, and Major Aashish had a six-year-old daughter. Major Aashish's mother bravely said, 'I refuse to cry. My son was a martyr who laid down his life in the service of the country.' Tear will come when you least expect them, grief will gut you and still you will continue to live.

Eventually the media would find another story; relatives would pack their bags and leave after the death ceremonies; and collective mourning would turn into individual pain congealed and sealed within you. Wives have lost their husbands, children their fathers, parents their sons, sisters their brothers, friends their comrades and defence and police forces their brave men. These martyred men will not see their children growing up and their children are too young to understand what they have lost. Who will hold their hands and help them cross this foggy bog of sorrow? And who will keep their memories alive? The war memorials with names engraved on stone walls are remembrances, ceremonially visited and forgotten till the next year.

There were tears flowing down my cheeks. Our brother's story was merging with theirs and I was not able to separate the two—where did one end and the other begin?

6

The Man

The real miracle is not to walk either on water
Or in thin air,
But to walk
On earth.

—Thich Nhat Hanh

After completing his tenure with the RR, Jeetu joined the NDA in the middle of January 2008 as an instructor on their training team. It must have been a matter of pride for him to return to his alma mater as an instructor. The gallantry awards were announced on 25 January 2008, with a lot of honours and accolades coming his way. After the investiture ceremony, Colonel Harjinder Singh as a caring senior, recollects cautioning Shatrujeet: 'Kotwal, at such a young age you have achieved much. Shaurat (fame) should not go to your head. Be careful about how you handle yourself.' Colonel Harjinder need not have worried; Jeetu was grounded with deep roots and the transient fame and glory would not have changed the kernel within. Biswajit, his squadron mate, remembers the personal influence Jeetu had on his life. 'I had congratulated Shatrujeet on the gallantry award and great guns his career was going. To this he had responded, "Thanks, Dash. But, I do not plan for long. Thinking too much about the future brings in anxiety. So, focus on today and also never fall into the trap of comparisons." This advice became the mantra for my life,' recalls Biswajit.

Jeetu fell into the rigours of his new assignment as instructor with full passion. I distinctly recollect receiving a call from him in the middle of March 2008. I was at work and my brother wanted to discuss neurolinguistic programming (NLP). In 2005, I had done a course—'Women in Leadership'—at the University of Bradford, UK in which Shakespeare's *Henry V* was used to teach leadership styles. During one of his annual leaves, Jeetu and I had discussed this particular play: the use of language, gestures and the 'royal face' that a king has to carry no matter what the crisis he is facing. Neurolinguistic programming is the practice of understanding how people organize their thinking, feeling, language and behaviour to produce the results they do. Formation of our unique mind maps of the world as a product of how we filter and perceive information absorbed through our five senses is the key element of NLP. Jeetu wanted to find the connection between NLP and my experience of learning leadership styles through *Henry V*.

Jeetu was always open to absorbing new ideas and also wanted to incorporate stories from the field in classroom teaching. Colonel Harjinder Singh remembers Jeetu discussing this in detail with him on joining the NDA, 'Sir, in the 34 RR we have done a number of good operations as a cohesive team in a tactical manner. There are a number of stories which we should include in training pedagogy; it will add immense value. Cadets would benefit from learnings of live operations'

Lieutenant Colonel Utsav, who was posted as adjutant, IMA, remembers: 'I was in the fifth semester and Shatrujeet Sir took classes on army tactics. Apart from the theoretical aspects, his lectures would be peppered with his experiences and stories from the valley. These "on-the-ground stories" resonated with the cadets and the classes were interesting.' The Lieutenant Colonel grins and goes on to add mischievously, 'Shatrujeet Sir was also liked by the cadets as he did not believe in giving "punishments" unlike some other instructors.'

I have quoted extensively from Jeetu's letter dated 25 February 1995 in Chapter 2 wherein he gave a day-by-day account of his SSB experience. In the same letter, while describing his interview as 'quite exciting and satisfying for me,' he also writes how the interviewer had asked him to name one quality in which he excelled and two qualities, which were his drawbacks. He wrote, 'I answered <u>confidence</u> for former and <u>sensitive</u> and <u>emotional</u> for the second one. Details afterwards.' (Underlining as in the original.)

True, Jeetu had all these qualities. He was confident, sensitive and emotional. Dear Brother, I want to assure you that being sensitive and emotional are not drawbacks, these are the qualities that made you the wonderful human being you were.

Abha and Jeetu married in March 2004 but could only stay together for barely one and half years—first at Ferozpur for a year or so and then six months at the NDA. The six months that they spent together in the NDA are imprinted in Abha's memory: 'Those were the best times. Shatrujeet was settling down after the hectic tenure with the 34 RR and was relaxed. I thought his health was improving and there was a glow on his face.' Vinkoo and her family had also travelled from Mumbai to visit Abha and Jeetu. Vinkoo recalls, 'It must have been February or March 2008 when we had gone to the NDA. Entry is restricted and when we were about to reach the entry gate, I had to call up Jeetu. He was waiting for us at the gate, standing next to his motorcycle. For a minute, I could not recognise him. I am not exaggerating but he was looking ethereal and there seemed to be some kind of "aura" around him.'

Vinkoo also remembers that trip to Pune with great fondness. 'Jeetu had taken us around the NDA campus and to the swimming pool too. Both of us climbed up to the highest diving board where he pretended to push me down as he would do when we were kids. We also went to their observation post somewhere high from where we saw such a beautiful night sky.'

For Abha too, those six months were the most beautiful time of their life together. 'He was fond of movies and for the first time in my life, I saw two movies back-to-back with him. We would go for long rides, have makki (corn) in the rain. We had bought bicycles and would cycle around NDA's beautiful campus.'

Abha was twenty-nine years old when Jeetu passed away; she never married again. With her voice heavy with unshed tears she says, 'We were planning for the future. Those six months were the happiest part of my life, my cup of joy was full. We had celebrated my birthday, a few days before the accident. What I had in this short time with Shatrujeet was enough to last for a lifetime. A whole life fast-forwarded and compressed in those few months of bliss. He was the most understanding partner one could ever have had.

'I am not an easy person to please and Shatrujeet has set standards that are difficult to match,' adds Abha wryly.

During our visit to Khadakwasla in June 2023, Abha was keen that we visit the house where she experienced the greatest of joys and the most excruciating of sorrows. Abha, Kartikeya, Vinkoo and I went to the house, where a lady lieutenant colonel lives now. The house is a small colonial-style bungalow with red tiles, a small veranda wrapped around it in the front, surrounded by huge trees. Looking at the house, I remember seeing a framed photograph of Jeetu in Vinkoo's house. He is in his uniform, sitting perched on the veranda railing with a luminous smile on his face. The picture was clicked by Vinkoo.

We walked around the house and I kept glancing at Abha to discern what was going through her head. What memories were flickering through her mind? Then she turned back and looked at us, pointing towards the window overlooking the veranda: 'This is where the drawing room is. It has two bedrooms with a study. It is a small, compact house. I remember hosting a few parties here.' Apart from these comments, Abha, stoic as ever, did not show any emotion. We took pictures with the house and then in silence got back into our car. On our drive back to Mumbai, all of us were lost in thoughts, in our private worlds.

During one of our conversations many years ago, Abha had mentioned that she had placed a request on a radio programme to dedicate Rabbi Shergill's Punjabi song 'Tere Bin' to Jeetu when he was posted in the 34 RR. She had hoped that he would listen to it and she would be able to convey her feelings to her partner via radio waves. Sometimes you find the right song, the right poem, the appropriate lines to describe your feelings. And Abha had found hers in the song. The first stanza, '*Tere Bin Sanu Sohaniya koi hornayo labhna, / Jo deve rooh nu sakun chuke so nakhra mera...*' It means, 'Without you my beloved there is no one else to be found who can give comfort to my soul and also bear my tantrums...'

Abha and Jeetu had a so-called Indian 'arranged marriage'. Both the families knew each other previously—photographs were exchanged, alliance were formed. Abha was born in Jammu but her family, like ours, also has roots in Bhadarwah. Abha's father, Satish Kotwal, who retired as chief engineer, was a deeply caring man with a stern appearance. He entertained no further discussion once a decision was taken. Their engagement was fixed in November 2003.

Abha had not yet met Jeetu though she had spoken to him over the phone a number of times. However, before the engagement, this bike-riding, speed aficionado who had completed her MBA from the University of Jammu insisted on meeting the man who was going to be her future husband. The grainy photograph of a serious, dark-complexioned Jeetu did not evoke too much confidence in her about her parents' choice of a life partner for her. She laughingly recalls that they met at KC Plaza in Jammu and her first impression on meeting him was, 'Oh! He is not as dark as he looks in the photograph. He is quite good looking.

'It was our first meeting and I was apprehensive. But he made me feel so relaxed. We talked nonstop on so many things and he was attentive, throughout. I never felt as if I was meeting him for the first time.'

Vinkoo also concurs, 'Jeetu had this innate ability to connect with everyone, whether it was neighbours, schoolmates, NDA or IMA batchmates, and later on army colleagues. He was equally comfortable with all relatives, irrespective of their age. And the best part was that he always tried communicating with them in a language they were comfortable with: Hindi, English, Bhadarwahi, Dogri or Punjabi. He used his sense of humour to make people around him feel at ease. He was a people's person, connecting with everyone around him, without carrying any prejudices or biases in his head.'

When I spoke to Ranbir, he became very emotional: '*Sahib ke saath aisa rishta ban gaya tha. Sahib Officer, mein sipahi.*' (I shared this unique connection with Sahib though he was an officer and I was a mere soldier).

Jeetu was a soulful, compassionate man who cared not only about human beings but also for other living creatures. Jeetu loved dogs and I am sure Abha and Jeetu would have found many common grounds in discussing motorcycles, cars and dogs, as Abha too is very fond of the four-legged furballs. Vinkoo remembers, 'I think he fell in love with dogs since the day our parents brought home a pup (an Alsatian) when Papa was posted in Delhi in 1981. He was named Jimmy after Jimmy Carter, the US president who had just lost the elections. After a few years, they brought another pup, Risky (an Apso), who became Jeetu's love. Unfortunately, Risky died in 1996, immediately after Papa's demise. Jeetu was in the NDA at that time. Once he joined the army, he adopted another dog, Shilpa, in Delhi.'

In fact, Shilpa was abandoned, and Jeetu had got her from a dog rescue shelter in Delhi. When he brought her to our home, she was weak, full of ticks, and barely able to move. He took her to Udhampur in the train, nursed her back to health with Mom and Arjun's assistance.

Shilpa accompanied Jeetu wherever he went and was a guest of honour at his engagement in 2003. When we travelled from Udhampur to Jammu, poor Shilpa had motion sickness, but Jeetu refused to travel without her. Shilpa and Jeetu were inseparable. In January 2005, when Shravani and I had gone to Ferozpur, Jeetu and Shilpa were at the railway station to receive us on that cold, foggy morning. They were devoted to one another and Shilpa would sense his arrival even before he entered the house. She would be all over him, licking his face, arms, and not leave him till Jeetu gave her a good scratch and played with her.

We lost Shilpa shortly after Jeetu's death—I think she was heartbroken. In 1888, the German philosopher Friedrich Nietzsche wrote in his book *Twilight of the Idols,* the aphorism: '*Aus der Kriegsschule des Lebens. Was mich nicht umbringt, macht mich stärker,*' which can be translated as 'Out of the war school of life, what doesn't kill me, makes me stronger.' He had written this in German, the language that Jeetu learnt at the NDA. Michael Brady, a professor of philosophy at the University of Glasgow, explains that Nietzsche does not seem to think that all suffering will result in strength, but rather that one should take suffering as an opportunity to build strength, and that those who are already strong are those who can do so. Jonathan Dollimore, an English philosopher, calls this process of embracing suffering as an opportunity to gain strength as a 'magical, cannibalistic ingestion'. We human beings are more resilient.

While reading a review of Naveen Kishore's book *Mother Muse Quintet* in the *Hindu Literary Magazine,* 7 April 2023, I was captivated by these lines quoted in the article: 'It is the seamlessness of remembering that amazes me. Memories. Yours and mine. The ones you enthralled me with on our long walks. And much later, my retelling of the "tales" you told me. "Making up" what had slipped past me.' Although the context is different—the poet is writing about his dead mother—the text voices my very thoughts.

Whenever he came to Delhi on leave, both of us would go for walks in the park in our colony and discuss family, career, politics, books that we had read, whatever came to our mind. He would walk unhurried. Once while walking, he had asked me, 'What is the length of this park in metres?'

'I do not know; I just make it a point to walk ten rounds,' I had replied.

'Okay, let us measure it now,' he had said with a shrug.

He had then proceeded to measure the distance with his footsteps and informed me it was around 300 metres.

This incident came to mind when I was trying to plumb my fallible memories. Jeetu was capable of having the deepest conversations and being goofy and hilarious at the same time.

Jeetu had a penchant for giving nicknames to everyone.

Since I wore spectacles, I had chaar (Hindi for four) eyes and was thus called 'Chucks'; Vinkoo was 'Billo' (billi is Hindi for cat) because of her light eyes; Arjun was 'Pindu'; Abha was 'Durku'; and Ranjeet was 'Scraggy'. After Jeetu graduated from the NDA, Mom was nicknamed 'Johnny Mom' based on Johnny Reb, who personified the ordinary Confederacy soldier in the American Civil War, as she was also going to be the mother of a soldier. Even Shravani was not spared: she was 'Latti' (simpleton in Bhadarwahi) because as a child she used to have difficulty in pronouncing certain letters of the alphabet and would trip over some of them.

The nicknames that Jeetu gave stuck. He never addressed me or Vinkoo as 'Didi' but only by the nicknames he had given us. Ranjeet still calls Mom 'Johnny Mom'. We also used Jeetu's skill to 'code-name' certain relatives and neighbours and would have a good laugh gossiping about them in our secret code language. 'During conversations, Jeetu's tongue-in-cheek comments sometimes went over the heads of some people but had us in splits,' remembers Vinkoo with a laugh.

Jeetu was a naughty child, full of pranks. Arjun idolized him and was his partner in crime. Holi was the perfect festival for both of them to trouble us all with their pichkaris (spray guns) and water balloons. *'Bhaiya ke jab 12th exam khatam hue tho hum* Dil *movie dekhne Chinar gaye the. Bahut masti ki thi, khana bhi bahar khaya tha'* (We had gone to Chinar to watch the movie *Dil* after

Bhaiya's 12th exam was over. Had a lot of fun and dined out), remembers Arjun. How fresh are Arjun's memories of the twenty-eight-year-old experience!

Food was another common love between Abha and Jeetu. Jeetu enjoyed food but unlike hardcore foodies was not obsessive about it. He enjoyed eating out. Abha recalls: 'During the six months that we were in Pune, I think we explored almost all the food joints around the NDA campus and Pune city. Pizza Point was a favourite. Both of us were fond of eating out, particularly non-vegetarian dishes.'

'I also recall Jeetu and Abha relishing the masala dosa and chicken biryani that I had made for them when they visited us in 2007, the year my family shifted to Mumbai from Delhi,' remembers Vinkoo.

'When we got married and went to Ferozpur, I did not know how to cook properly. Mom was there and she imparted the basic cooking skills. I then asked Shatrujeet to buy a mixer-grinder from the army canteen. Instead of buying a simple mixer-grinder, he bought a food processor and told me, it will be useful not only for cutting and grinding but for kneading flour too. He was empathetic and cared.

'Though Shatrujeet enjoyed eating out, he was equally appreciative of what was cooked at home and I do not recall him ever being critical of what was cooked and placed on the dining table to eat. Once when I had made kachumber salad (finely diced cucumber, tomatoes, onion and chillies tossed with a hint of lemon) at Ferozpur, both Shatrujeet and Mom praised it so much as if I had prepared a gourmet dish. Both of them were encouraging as I took baby steps to learn cooking.' Under Mom's guidance and Jeetu's encouragement, Abha became a pro in a few months. By the end of her stay at Ferozpur, she had turned into a wonderful hostess, serving delicious Kashmiri aloo-dum and mutton curry at dinner parties.

'Shatrujeet also liked his drink, particularly whisky and beer,' recalls Lieutenant Colonel Dass. However, he was careful of how much he drank. Colonel Bhola remembers: 'Shatrujeet had come to Pune for a course in 2005 and I was posted there. He came home to dinner and we had a few drinks. I offered him some more and he refused, saying, "Bhola, the aim of drinking is to enjoy your one or two drinks; not to get drunk."' Colonel Harjinder echoed the same thoughts: 'On his dining out from the 34 RR, he was being plied with drinks, but after some time he politely refused. That boy knew not only how to carry his drink but also when to say no, which is very important.'

I am the eldest of three siblings and was a disciplinarian. I supervised Vinkoo and Jeetu's studies till they were in the eighth or ninth class and insisted on them getting good grades. During power cuts, which were quite frequent in Kotli Pain, we would sit on the lawn and play a game of antakshari but not with songs. We would do that with countries and their capitals. It was my way of brushing up their general knowledge. I would also assign them domestic work around the house. After meals, it was Jeetu's duty to clear the dining table and fill water bottles to keep in the refrigerator, while Vinkoo and I helped Mom in the kitchen. I would insist that they keep the almirahs tidy too. I am sure both of them hated me for being such an ogre.

Years later, in our home in Delhi, when I would be cooking in the kitchen, Jeetu would hang around, talking and drinking his tea. Then suddenly, I would ask him to get something from the refrigerator and he would tell Madhu cheekily, 'Bhaiya, never be in Chucks' orbit. She is a workaholic, always ready to boss around and assign you work.' Madhu would nod his head vigorously in agreement.

We were blessed to have a brother like Jeetu in our lives.

Papa's passing away had a profound effect on Jeetu. Kshitij also remembers, 'After his father's death in the second-term, Shatrujeet did not show any overt emotions but for a long-time, he kept to himself.'

'He was not even nineteen when Papa passed away. He was very obedient and responsible from the beginning, but I guess he became more responsible towards us, his sisters and Mom, after Papa's demise,' observes Vinkoo. Dheeraj also echoes similar thoughts, 'Shatrujeet was very concerned and protective about the three of you, particularly about his younger sister. He would tell me, "I am not worried about Didi, but Vinkoo would need me around her."'

Yes, brother, we need you around us.

Once Jeetu started earning, he made it a point to give us very thoughtful gifts. Vinkoo recalls, 'I remember a very beautiful set of pearls that he gifted to me very early in his career. From Mhow, he got us an enchanting painting that we still have today. He always got gifts for Abha, Mom and Vinoo didi.'

Yes, indeed he did. I still have the beautiful Paithani that he got from Pune.

Jeetu himself was not too particular about what he wore, and I think his dress sense improved after Abha started buying clothes for him! However, he appreciated what others wore. Colonel Harjinder remembers with a chuckle, 'I am fond of wearing smart shirts and have a good collection. Both Shatrujeet and Raghu would appreciate my shirts and ask quizzically, "*Sir, kahan se laate ho itni achi shirts?*" (Sir, from where do you bring such nice shirts?).'

Jeetu was, however, very fond of his leather jacket that he had bought from Chandigarh when he was posted in Solan. Riding his motorbike, wearing the leather jacket with his Ray-Ban aviators, he looked like Tom Cruise, his *Top Gun* hero.

Jeetu was a caring brother. On our visit to Ferozpur in January 2005, Jeetu and Abha were the perfect hosts, taking us to the Golden Temple and Jallianwala Bagh in Amritsar in their Maruti Zen car. It was an exciting trip. It had rained the previous night and the car got stuck in mud at one point on the road. With Abha at the wheel,

Jeetu and I pushing from the back, and got the car out with great difficulty.

The Golden Temple, also known as Sri Harmandir Sahib, is one of the holiest of the sites of Sikhism. The holy golden sanctum inside the pool is a sight to behold. A sense of peace descends on you while doing the parikrama (circumambulation) and listening to the mellifluous shabad (chanting) suffusing the air. I was reminded of Guru Nanak's philosophy of *Ek Onkar Satnam, Karta Purakh, Nirbhau, Nirvair, Akal Murat, Ajuni Saibhang, Gur Prasad* (God is but one, his name is true, he is the creator, he pervades the whole universe, he is without fear, he is without enmity, he is immortal, he is birth-less, he is self-existent, he is remover of darkness [ignorance] and he is merciful). How powerful and profound are these thoughts!

We also had the Guru ka Langar, a free meal in the gurdwara's community hall. We sat on the floor and were served dal on the plate; if one wants a roti, one has to raise one's hands to accept it with religiosity and gratitude. Before beginning to eat, one has to respond to the call of 'Jo Bole So Nihal' (Whoever utters [the holy name] shall be happy), with 'Sat Sri Akal'(True is the name of God). It was one of the tastiest meals I have ever had. Shravani was fascinated with the automatic roti-maker installed in the langar kitchen.

From the Golden Temple, we went to Jallianwala Bagh, the site of the Amritsar massacre of 13 April 1919, the day of the traditional festival of Baisakhi. The seven-acre site now houses a museum, a gallery and several memorial structures, the most important being the Martyrs Well, which is surrounded by the Martyr's Memorial, a large structure with a sign displaying '120'—the number of bodies that were recovered from there. All of us were pensive, looking at the wall pockmarked with bullets. The road to India's independence was an arduous journey with sacrifices of so many unsung heroes.

We also went to the Hussainiwala border, eleven kilometres from Ferozpur, to watch the Beating the Retreat ceremony to mark the lowering of the national flags of India and Pakistan with full honour and respect every evening. This checkpoint is about thirty metres from the Zero Line on the Indian side and a hundred and eighty metres from the Pakistan side. About 4.5 metres from the line on either side, the Indian and Pakistan national flags are hoisted during the day. This joint parade and the retreat ceremony were started in 1970 and has become a tradition now. As the sun sets, the silence is broken by the clattering and thundering of heavy boots of the jawans, smartly dressed in the khaki uniforms of India's Border Security Force. On the other side, giving them fierce competition, are their counterparts, the Pakistani Rangers. Both the troops move in unison, performing identical movements in step with each other, under loud commands given by the respective picket commanders, throughout the forty-minute retreat ceremony. Shravani sat mesmerized watching the ceremony, which intensified the feelings of patriotism in the audience.

We had also visited the memorials of Shahid Bhagat Singh, Rajguru and Sukhdev, which are just a kilometre away on the Indian side. This trip to Ferozpur remains fresh in my mind. It was lovely to see Abha and Jeetu in their happy domesticity, with Shilpa in tow. Abha and Jeetu gifted me a beautiful phulkari dupatta that I still wear. The trip was full of adventure, joy and togetherness.

Of the three siblings, Jeetu was the youngest. However, he was aware of the frailties of human relationships and how to maintain them. During one of his annual vacations, Jeetu, along with Ranjeet and his other school friends, including Naveen Suri had planned a roadtrip to Dharamshala and other surrounding areas in Himachal Pradesh. It was around Diwali in 2002, just before we arrived to celebrate Diwali at Udhampur. Years later, Ranjeet told me that Jeetu had cut the trip short to be back home before we reached. 'Chucks and Jijaji are coming to Udhampur to celebrate Diwali and I should be there to receive them.'

'Shatrujeet was a great guy to have as a friend. His thought process was way ahead of his counterparts. In June 2008, around the Passing-out Parade Day, I went with my wife to Shatrujeet's residence for breakfast. Over hot aloo parathas, we had these deep, incisive conversations on international affairs as Shatrujeet had been selected for the UN mission in Congo and was due to leave in July. Then suddenly, we changed gears and started discussing the army as a career. It was good fun, sparring with him over different topics,' reminisces Lieutenant Colonel Dass. He continues, 'I lost my elder brother long back, when I was very young. Time passes, people move on. Memories remain. Some memories of the time spent with Shatrujeet are embedded in my mind, forever.'

Speaking to his seniors, colleagues, friends, and subordinates, I can picture Jeetu as a fun, dependable, and compassionate human being. These qualities spilled over into his personal life too. 'Jeetu always displayed equanimity in his thinking and behaviour. He was no doubt very different from others, from us. Few people reach this stage of equanimity or balance and many of us don't even reach it. He just came, saw, and conquered the real meaning of life only in those brief thirty years that he spent with us,' concludes Vinkoo.

7

Death Came Knocking

I nursed it in my bosom while it lived,
I hid it in my heart when it was dead;
In joy I sat alone, even so I grieved
Alone and nothing said.

—Christina Rossetti
'Memory'

On 17 June 2008, I received the news of Jeetu's death on my mobile phone while I was returning home from office on the Delhi Transport Corporation (DTC) bus 604. It must have been around 6.45-7 p.m. The caller was Colonel Chauhan from the NDA: 'Good evening, Ma'am. I am Colonel Chauhan from the NDA, Khadakwasla. Am I speaking to Major Shatrujeet Kotwal's sister?'

I was a little taken aback at receiving the call and answered in the affirmative.

'Ma'am, I have bad news for you. Major Kotwal died

in a road accident this evening. My heartfelt condolences.'

I do not recall the rest of the conversation, how I got down from the bus, crossed the main road and went to my friend Vasanti's house from where I was to collect Shravani. Everything was a blur.

Cheryl Strayed, the author of *Wild*, likens grief to a box you'll hold forever: what it means to you is determined by how bravely you look into it. I am still holding that box with these fragments that I keep looking at.

We left for Jammu the next morning by an early-morning flight, tickets for which had been arranged by Vasanti and her husband, Prem. They drove us to the airport. That night is seared in my mind. I remember pacing up and down the whole night with tears streaming down my cheeks, not able to make sense of what had happened. I could not breathe. Shravani was barely eight and half years old and was traumatized on seeing what was happening around her. For a long time, her eyes would well up whenever Jeetu's name was mentioned. With age, she has also learned to process and make sense of the grief.

Another fragment in my grief box is meeting my mother after Jeetu's death. She was sitting there in the drawing room on the ground surrounded by wailing women. She looked at me and said, '*Beta woh hame chod kar chala gaya*' (He has left us and gone) and started sobbing inconsolably. We held each other, our tears mingling. Scenes from the homes of Colonel Manjeet Singh, Major Ashish and DSP Himanyun reminded me of that day again. When I saw that evening news in 2023, like a cinema reel in a loop, the memories of 2008 started playing in my head.

A young man with the world at his feet was gone forever. We would not be able to see him, talk to him, touch him, hug him or laugh with him. How final is death. We take the body to the cremation ground, light the pyre and everything is over. I also find it bewildering that the transition from a living person to a body is so swift and brutal. The one who was so dear to us suddenly becomes untouchable and we gingerly go around the body, not allowed to touch it.

On 18 June 2008, Jeetu's embalmed body was flown in an Indian Air Force plane from Pune to the Udhampur Air Force Station and taken straight to the Military Base Hospital. At the air force station, Madhu and I received Jeetu in a wooden coffin. I do not recall who else was there with us, but I remember that there was an officer accompanying his body from Pune. I remember clinging to that wooden box, trying to hug my brother and crying inconsolably on our journey in the three-tonne truck to the Base Hospital from the station. I refused to let go and was forcibly separated from him. There was this white blaze of anger inside me. From a haze of pain and anger, I could see many moist eyes.

Abha, Vinkoo and her family came from Pune to Delhi on a domestic flight since civilians could not travel in an Indian Air Force plane. From Delhi, they drove down to Udhampur—a five-hundred-kilometre journey through Punjab and Haryana. My sister has a distinct recollection of the journey: 'It was the most difficult and painful journey I had ever taken. We are grieving and at the same time trying to be brave too. While travelling through Punjab in the night, we lost our way. Anadi was barely four years old and I remember him clinging to me throughout the journey. By the time we reached Udhampur, it was almost the morning of 19 June.'

I hugged Abha and she whispered to me, '*Uff, yeh kya ho gaya* (What is this that has happened).'

They did not allow us to open the coffin when we arrived at the base hospital. It was only opened the next day on 19 June when we came to take him for his final journey, from the home where we had dreamt, aspired, and spent our days of joy and sorrow. That squat concrete building was a repository of our collective memories, and Jeetu came to the same house to undertake his final journey—the same journey that, fifteen years ago, our father had undertaken from the same place. We were not allowed to touch the body: we could only look at Jeetu, who was there and yet not there.

He was cremated with full military honours befiting a hero on the bank of the river Devika. According to the Nilmat Puran, written in the seventh century by Nela Muni, the Devika river is a manifestation of the Mother Goddess Parvati herself. To benefit the people of Mader Desha, covering the areas between the rivers Ravi and Chenab, the Devika appeared on Shivratri. Lord Shiva himself remains by the side of his consort Uma, manifesting himself in the form of Shiva lingas at eight places along the river. It is believed that being cremated on the sands of the Devika is as good as being cremated on the banks of the Ganges at Kashi.

There were hundreds on the bank of Devika to bid farewell to Jeetu. Abha lit the pyre.

Mom had not come to the cremation ground. She could not bear to see her youngest consigned to flames. When Jeetu was young, Mom would sing a lullaby written by Padma Sachdev, a progressive Dogri poetess and writer who died in 2021. *'Tu malla tu, Log panan thi kriyan, Badam panne tu... Tu meri umar bhi bhog bachua, Tere kola nalange koi rog bachua. Ae me din gini gini leta ae, ae me sukhan sarini leta ae...'*

A mother is lulling her child to sleep and singing softly, 'My dear child, may you always break almonds while others break flakes of broken pots. No illness should ever go near you, you may live my life too. I have got you after so many days of prayers and grace...'

Listening to this lullaby in Lata Mangeshkar's voice always gives me goosebumps and I can vividly see Mom sleeping sideways on the bed and patting Jeetu to sleep. How could she have gone to the cremation ground?

There can be nothing more traumatic for parents than the death of their child.

The cremation was widely reported in the local newspapers, particularly the fact that Abha had lit the pyre.

Later, Abha and Ramprakash Chachu (one of my father's younger brothers) would complete the daily ceremonies till the thirteenth day. The entire extended family was distraught with grief and at the same time bewildered. Everyone kept nodding their heads in disbelief, repeating as if in a loop, 'How could this have happened? Why him? He was so young...'

There were no answers.

After the cremation, Abha along with Ramprakash Chachu performed the shradh ceremonies. In a strange way, these ceremonies not only mark the end of life but also prepare for transition to another life. The offering of small rice balls (pinda pradana) during the rituals are thought to feed the deceased prior to his or her rebirth and a lamp is lit giving hope to the soul that it has not been forgotten. During these days, there was a constant movement of people, including friends and relatives. This collective sharing of grief is meant to ease the burden of loss.

In July 2008, we took Jeetu's ashes to Haridwar for the Asthi-visarjan (the immersion in the river Ganga) or the *nirava panjali*. The physical or the material body is considered to be made of the *paanch tatva* (five elements of Earth, Air, Water, Fire and Ether). During cremation, the body merges with the five elements, and the gases and other compounds present in the body are freed, which then merge into the atmosphere. Only a few bones or ashes (ashti) remain and these have to find a resting place.

Hence, the Asthi-visarjan—so that the soul may finally rise to heaven.

Legend has it that King Bhagirath had performed a tapasya (penance) to bring the river Ganga down from the heaven to Earth so that the souls of sixty thousand of his slain ancestors dwelling in the underworld could attain nirvana and be released when touched by the river of heaven. After many years of tapasya, Bhagirath could finally appease Ganga who agreed to descend on Earth. Shiva decided to channelize her flow through his matted locks. Bhagirath led Ganga across the Earth, then to the sea and finally to the underworld where the ashes of his ancestors were purified by her holy waters, and they were finally allowed entry to the other world.

Mom and Abha accompanied by Abha's parents, Deepa, Abha's sister, and Pawan Kotwal, Abha's brother-in-law, travelled from Jammu to Haridwar. Madhu and I along with Rekha didi, our first cousin, came from Delhi. Haridwar has traditionally been the site for performing the last rites of Hindus, and the family pandits or pandas who maintain the genealogy registers of families reside there. Over the centuries, these voluminous handwritten registers are passed down through the generations by their pandit ancestors. I was quite surprised to see that these registers are classified according to the original districts and villages of one's ancestors. Our family panda also took our register and there I could see the names of my father, grandfather, great-grandfather, grandmother, uncles...We were adding Jeetu to the list.

I often scribble quotes, poems, facts etc. in my diary. On 17 June 2008, I wrote 'Life changed forever'; on the facing page appear the following lines from 'Life and Death' by J. Krishnamurti:

Death must be something
Extraordinary, as life is
Life is a total thing
Sorrow, pain, anguish, joy,
Absurd ideas, possession,
Envy, love, the aching
Misery of loneliness—
All that is life. And to
Understand death,
We must understand the whole life,
Not take just one fragment of it and live
With that fragment, as most of us do. In the
Very understanding of life there is
Understanding of life, there is
Understanding of death, because the two
Are not separate.

I am sure I must have reproduced these lines in my diary after Jeetu's death but I do not remember writing them. Still, these lines are there and how true they are. We start dying the day we are born. The inextricability of life and death and our ability to cope have been the subject of many theological texts, academic tomes, popular books. Sir Thomas Browne in his book *Religio Medici* (The Religion of a Doctor) notes, 'With what strife and pain we come into the world we know not, but 'tis commonly no easy matter to get out of it.' Jeetu died in a road accident, and I am haunted by this thought hoping that our brother did not suffer for too long in his last moments.

Many years later, reading Paul Kalanithi's achingly beautiful book *When Breath Becomes Air*, which is dedicated to his daughter Cady, I made peace with this thought too. Paul was diagnosed with stage IV metastatic lung cancer when he was thirty-six and on the verge of completing his intensive decade-long training to become a neurosurgeon. However, life, destiny—whatever you may call it—had different plans. Within two years of his diagnosis and just a month short of his thirty-eighth birthday he passed away. Between the diagnosis and his death, he managed to gift us this wise book. Reading this book made me feel the pain and suffering that he underwent during the repeated cycles of chemotherapy. He bore all with courage, dignity and grace, and I am sure our brother did the same. Incidentally, Paul was born in April 1977 and Jeetu was born in August of the same year.

With Jeetu gone, all of us tried picking up the pieces of our altered lives and making sense of it. It was not—and it still is not—easy. Abha tells me, 'Sometimes while driving, I used to shut my car windows, ramp up the music and scream.' I can imagine a twenty-nine-year-old, trying to be brave before our aging parents and relatives, yet needing to vent her unbearable, choking grief. Jayson Greene in his memoir *Once More We Saw Stars* writes about the loss of his two-year-old daughter, Greta. One minute, Greta is sitting on a bench with her grandmother outside their building in the Upper Side of Manhattan: the next minute, a crumbling brick from the windowsill overhead strikes her. She is rushed to the hospital unconscious, where she dies. How do you explain this? In this book about grief and healing, hope and despair, Greene writes about 'prospecting' for safe screaming spaces in New York while his wife Stacy does what Abha did.

After our trip to Haridwar, towards end of July 2008, I received a call from Mom. There was a strange timbre in her voice and then wobbily, she told me, 'Abha is pregnant'. I was overwhelmed with emotions. What is happening! Barely a month ago, we had lost our brother, unaware that Abha was carrying his child. Too much was happening. I really do not recall whether I spoke to Abha immediately or after some time. But, this incessant voice in my head kept repeating in a loop, 'What will Abha do? How will she bear one shock after the other?' Everything passed in a haze and Jeetu's son, was born in February 2009. Vinkoo flew from Mumbai to Jammu for Abha's delivery and I went for the baarhol, the twelfth-day ceremony that is performed for the new-born.

8

Possibilities

Everything is attainable,
Possibilities unfurl like new lime-green leaves,
Some are singed by the harsh sun,
Others adapt like the Moon cactus,
Grafted together, they flourish.

Abha named their son Kartikeya. I was a little intrigued by her choice. Kartikeya, the son of Shiva and Parvati, is also known by many names—Murugan, Shanmukha (Shanmuga), Guha, Senthil, etc. in south of Vindhyas. He is the God of War and Victory and this name is more pouplar there. I asked her about it and she looked me straight in the eye and said, 'Shatrujeet, asked me.' Looking at the bewildered expression on my face, she said, 'During preganancy, once Shatrujeet came in my dream. He was standing there tall and staright in front of me, looking as hansome as ever. His eyes were shining so brightly that I could barely look at him. He told me if the child is a boy, name him Kartikeya and if it's a girl name her...

'I woke up startled and Deepa didi who was sleeping next me, also got up worried. I told her about Shatrujeet coming in the dream and teling me about the names for the child. But, strangely I could not recall the name for the girl he had suggested,' recalls Abha with a faraway look on her face.

A warrior choosing a warrior name for his son seems to be appropriate. Not only is Kartikeya a god of combat and triumph but also the epitome of intelligence and perfection. After the rituals and visit to Haridwar, Abha went back to her parents' home in Jammu. Apart from her parents, her sister and family lived nearby. During Abha's pregnancy, we would be in regular touch with her but it is Deepa, her sister, who stood beside her like a rock. She would accompany Abha for the check-ups with the doctor and tend to her like a child. 'She was like my second mother,' remembers Abha. Deepa gave her the physical, emotional and spiritual support through these difficult times and continues to give it. Mom would also go from Udhampur to meet her. Though Abha and her unborn child were surrounded by this envelope of love, I am sure she must have missed her husband. I could never gather the courage to broach this subject with her. It was like walking on eggshells, lest the cloak of normalcy disintegrates.

Now, so many years later, during my conversations

with Abha for the book, I finally opened the doors that had been kept firmly locked. How did she feel when she discovered that she was pregnant? Did this overwhelm her as she was already in such a fragile state? Why did she not get married again?

Abha looked at me quizzically and asks, 'Aren't you a little late in asking these questions? How does this matter now?' I gently remind her that we have had very brief conversations on the subject in the past, just skimming over and, I admit, never in great depth. Abha would always tell me, 'I will not meet anyone better than Shatrujeet.'

Years later, she opens up and tells me, 'When I got married, I had male counsins who were in the same age-group or slightly younger than Shatrujeet. After meeting Shatrujeet, I realized, the difference was striking. Here was this man so young but with so much of thehrav (calmness and composure) in his demeanor that others paled in comparison. For one so young, he was deeply spiritual, an old soul. I am a very difficult person to please and I do not think after being married to Shatrujeet, I would have been happy with anyone else.'

She turns serious and says, 'Through Shatrujeet, you people came into my life—you, Vinkoo didi and Mom. Apart from my parents, Deepa didi, Pawan jiju, my extended family, you people have been there for me and I have always felt surrounded by love and warmth. I know all of you are there for me and Kartikeya, no matter what. Why would I give up on you people? I am very possessive about the people I love. I do not want to give this up for something unknown.

'We were married in March 2004. Our marriage was just a little over four years old. Out of these, we were barely together for one and half years. I was ecstatic during our stay in Pune and then one day, he just disasppeared from my life. He just slipped from my hands and I did not know what was happening. It was like a free fall into abyss. And then this miracle happens and I have a new life growing inside me. How could I not be part of this miracle for all of us?'

As Ocean Vuong poignantly says in the poem 'Headfirst':

> But only a mother can walk, with the weight, of second beating heart.

Abha continues somberly, 'I had seen how all of you were devastated with grief and sorrow. This new life gave hope and solace to all of us. How could I not be the instrument for this happiness? Having Kartikeya in my life has been the best decision. As he grows up, I also evolve and together we build this sanctuary of hope, love and contentment.'

She continues, 'Having Shatrujeet in my life was meant to be. When I met him for the first time, I was so comfortable being with him, it never felt as if it was the first time. Who knows, maybe I have this karmic connection with Shatrujeet and we are bound together? Sometimes, life is too complex to be described in words! We should not seek explanantions or look for answers for everything.'

9

Stages of Grief

Five or Seven?

I measure every Grief I meet
With narrow, probing, eyes-
I wonder if It weighs like Mine-
Or has an Easier size.

—Emily Dickinson
(Composed 1862, published 1896)

Based on her years of working with terminally ill individuals, the Swiss-American psychiatrist Elizabeth Kubler-Ross wrote in her book *On Death and Dying* that grief could be divided into five stages: denial, anger, bargaining, depression, and acceptance. The Kubler-Ross model, though originally designed for people who were ill, has been adopted to deal with death, loss etc.

In another system, grief has seven stages, the five stages as proposed by Kubler-Ross being supplemented with two more, namely guilt and reconstruction, an upward turn separating the two. Then, there is the four-step model, the ten-step model...

Grief is universal, but your grief is different from mine. Jeetu's sudden death left us grieving beyond words. My mother had lost her son; Abha, her husband; Vinkoo and I, our brother; Shravani and Anadi, their Mamu (maternal uncle); Madhu and Pabitra (Vinkoo's husband), their brother-in-law; and yet others, their son-in-law, friend, nephew—and Abha's then unborn son had lost his father. Jeetu meant so much to so many different people. We were bound in our grief together and yet apart. We hardly spoke about it. Years later, speaking to Dheeraj, I came to know that Jeetu still appears in his dreams. 'I always see him in his formal pink shirt that suited him well and I repeat the question, "Where were you? Did not see you for long." And he replies smilingly, "I was badly hurt, and not well. But, now I am fully recovered, do not worry."'

Dheeraj ruefully shrugs, 'I do not dream so much about Pankaj.' These three buddies from the NDA had forged such a strong connection transcending the conventional blood ties, and now one of them still remembers the other two with such intensity.

In June 2009, we had gathered again in Udhampur for the annual death anniversary (colloquially called 'barakhi'), the date for which is calculated based on the tithi, the exact date and time of death. After the puja and yagna, there is an offering (daan) made to the priests of all the things that a soul may require for her/his journey to the other world. Starting from shoes, umbrella, utensils, bed, mattresses, etc., everything is given. The previous year, for the thirteenth-day ceremony, I had gone to the main Udhampur market and bought these things for my brother's final journey. Ramprakash Chachu had accompanied me and choking back our tears, we had selected and bought all the things.

For the barakhi, all the stuff for daan were displayed on the cot outside the veranda when it suddenly started pouring heavily. Although the rain only lasted five or ten minutes, the fact that it happened on a clear, sunny day, without a wisp of cloud in sight, stunned everyone. Like we humans always do, we had to offer explanations. Someone explained that the gods had blessed the ceremony. However, I do not know how to make sense of this freak weather phenomenon. And does conducting a barakhi mean that life moves on as if everything is normal now?

Lisa Marie Presley, Elvis Presley's daughter, who died on 12 January 2023, wrote this in *People's Magazine* (2022) about her son, Benjamin Keough, who had killed himself in 2020: 'Grief is something you will have to carry with you for the rest of your life, inspite of what certain people or our cultures want us to believe. You do not "get over it", you do not "move on", period.'

I can understand what she is saying. A fragment of a song, a glimpse of a young officer in his camouflage riding a bike, can break the levee...Even a picture appearing in the obituary section of a newspaper can unhinge you for a minute as I did, recalling the poem 'Obituary' in my book of poems, *Twilight Wanderings*:

> I scan through the obituaries,
> Very carefully,
> Walking through a minefield,
> Particularly, looking at photographs,
> Of men in Uniforms.
> Mental maths,
> At which I was never good,
> I calculate their ages.
> Brother,
> Some are younger than you,
> And some older.

We passed through these so-called stages of grief, coping and handling them in our own personal ways. Personally, I immersed myself in my work. There was a career transition for me in 2008, and I used this challenge to work nonstop. For me to be in motion, constant activity was a distraction. I would not discuss Jeetu's death with anyone but my search for answers on what had happened continued. I have always found comfort and succour in words. Different people have their distinct mechanisms for dealing with stress or searching for answers—some sleep, some talk with friends, others go for a run and I turn towards books.

People might find it strange but I have always felt that books not only speak to you but listen too. I tried to read as much as I could to make sense of what had just happened. I read *Year of Magical Thinking* and *Blue Nights* by Joan Didion, *Paula* by Isabel Allende, *The Last Lecture* by Randy Pausch, *When Bad Things Happen to Good People* by Rabbi Harold S. Kushner, *Oona* by Jasjit Mansingh back-to-back, cover to cover. All books about the loss of loved ones.

On Christmas day 2003, Didion's daughter Quintana, who had come down with flu-like symptoms, went to the emergency room suffering from pneumonia and septic shock. And then she was suddenly in the hospital's intensive-care unit, hooked up to a respirator. Five days later, Didion's husband of forty years, John Gregory Dunne, died of a massive heart attack at the dining table. The *Year of Magical Thinking* follows Didion reliving and re-analysing her husband's death throughout the year that followed. 'Life changes fast,' writes Didion. 'Life changes in the instant. You sit down to dinner and life as you know it ends.'

How true! That one call on 17 June 2008 changed our lives forever. Now, everything is benchmarked in my head with reference to this day and year.

Joan Didion's daughter Quintana survived her visit to the emergency room in 2003 but died of pancreatitis in August 2005, aged thirty-nine, before the *Year of Magical Thinking* was published. Didion wrote her second book, *Blue Nights*, after her daughter's death. These two books chronicle her heartbreaking losses.

Paula, by Isabel Allende, the Chilean author, is about her coping with her daughter going into a coma, which left her in a vegetative state, and then dying a year later, at the age of twenty-nine. *Oona: Mountain Wind,* another book by a mother, Jasjit Mansingh, is about Oona, her thirty-four-year-old daughter, who along with her three-and-a-half-year-old daughter Ilya (Jasjit's granddaughter), died of mushroom poisoning. In book after book, I searched for answers.

'Life is not fair', writes Rabbi Harold S. Kushner in *When Bad Things Happen to Good People,* and none of us can avoid the problem of why bad things happen to good people. However, what is important is how we respond, what we intend to do, now that it has happened.

Years later, I read Patrick O' Malley's blog 'Getting Grief Right' (2015), where he talks about chapters rather than stages of grief. According to him, Chapter 1 has to do with attachment, the strength of the bond with the person who has been lost. According to him, the size of grief corresponds to the depth of their love.

Chapter 2 is the death event itself. At this stage you question your sanity.

Chapter 3 is the long road that begins when the outside world stops grieving with you.

He also writes: 'The truth is that grief is as unique as a fingerprint, conforms to no timetable or societal expectation. When loss is a story, there is no right or wrong way to grieve. There is no pressure to move on.

There is no shame in intensity or duration. Sadness, regret, confusion, yearning and all the experiences of grief become part of the narrative of love for the one who died.'

Since Jeetu's death, I have been grappling with questions about human mortality. Greta (2); Aaron (14); Paula (29); Oona (34) and Ilya (3½); Quintana (39)... These are the lives about whom I have read. What about those who are around us and leave suddenly? On 6 June 2022, we heard that Sheil Sagar, Shravani's twenty-two-year-old classmate and friend, was no more. That beautiful young man with a curly mop of hair and soulful voice was suddenly just gone. Every time I hear Sheil's song 'If I tried' on Spotify, I cry. I wonder what his parents must be going through. Why did he have to go so young? What is the rationale of all of this? What is the pattern? Why them?

There are no answers and now I understand making sense is futile. We live through loss and try to clear the brambles of grief and sorrow with a machete. Viktor Frankl, the creator of the Logo therapy, in his book *Man's Search for Meaning*, says, 'Suffering ceases to be suffering at the moment it finds a meaning. Meaning is something that we discover rather than invent. We have to find it ourselves. We find it through living, and specifically through love, creating things and the way we choose to see things.'

It would, however, help if we could navigate these difficult times with some help, particularly when the outside world stops grieving with you. While reading Jayson Greene's book, what struck me most was the fact that they dealt with the loss of their daughter through their combined individual strength as well as external support they sought through 'grief counseling'.

Shilpi, my friend, had recently lost her father in England after a brief illness. While expressing my condolences, we got talking about aging parents and our own mortality. She said, 'After my father's death, we had a bereavement counsellor come over to speak to my mother and the rest of us. This support is invaluable at such a difficult time.' The gov.uk website also has information on what to do when someone dies.

In India, our immediate and extended families were meant to provide this support. The twelve-to-thirteen-day mourning period was the first bulwark against the storm and thereafter the monthly ceremonies for the deceased. With society in a flux, disintegrating old structures and new ones being re-imagined and re-worked, we do not have certainty of this support any more. Nuclear families in cities with both partners working find it difficult to cope with competing pressures.

Till a few years back, talking about mental health was taboo, especially in our society. Only weak people indulge in crying; being sad and acknowledging that one is feeling sad and under the weather was unimaginable. We are supposed to present this perfect image of being happy all the time. The *TIME* magazine in its 14 August

2023 issue profiled John Fetterman, a US, Senator, who after surviving a stroke and winning the most expensive US senate race ever, fought depression, went for treatment and is now talking about it in order to dispel the stigma attached with depression and other mental health issues. In the article, the writer Molly Ball writes, 'For so many years, we have demanded our politicians be perfect—free of scandal, perfectly groomed, never a hair or a word out of place. To admit to being broken was to admit being deficient.'

The recent incident involving Colonel Manjeet Singh and others made me think about these lines again in another context. On 15 September 2023, the electronic media went into a hyperbolic state once again, flashing visuals of six-year-old Kabir Singh, son of Colonel Manjeet Singh, dressed in army camouflage, saluting his father's coffin. Why do we want to portray this picture of hyper masculinity? Is it not unfair to expose the child to such trauma? On top of this, we have these news anchors loudly proclaiming, a martyr's son pays homage to his father—this is the young generation of future India! The line between martyrdom and patriotism—can we not discern between them? Can one not be a patriot without being a martyr? Why does a six year old have to be burdened with his father's legacy? Why should he not be allowed to forge his own path?

Why do we not let the family grieve in peace and dignity, without making a spectacle of their immeasurable loss? What is needed after all the hullabaloo is over and the camera lights are switched off is grief counselling for the family. Difficult times lie ahead as they grapple with this new reality. Children will grow up asking questions about their father; most of the time spouses will bear the responsibility of being a single parent; and old parents would need emotional support. Sometimes speaking to an emphatic stranger equipped to address such issues can be cathartic.

Epilogue

Lucy Kalanithi, wife of Paul Kalanithi, in the epilogue of *When Breath Becomes Air*, writes, 'What happened to Paul was tragic, but he was not a tragedy.'

I want to say the same thing for Jeetu; yet, no sister should be doing what I am doing, writing about her dead brother. This story should have been written by Jeetu himself as I see so many ex-defence officers penning their memoirs with great flourish.

The last time I met Jeetu was towards the end of May 2008 when he had come to Delhi for his interview for the UN assignment. He was selected and was due to leave for Congo in July 2008 after his training at the International UN Military Observer Course (UMMOC-8). He was so excited to embark on this phase of his career. He was fond of travelling and would ply me with a thousand questions on my trips abroad. He wanted to see Europe. I look at his batchmates with their families, doing well in their career; I am happy and wish them well. But why was all this denied to Jeetu and his family? Why was a life so full of possibilities cut short so brutally? His posting as an instructor at the NDA was one of short respite, before he embarked on another challenging posting. He had reached this pinnacle in his short career through sheer dint of his hard-work, perserverance and pushing the physical as well as mental boundaries. Loosing Papa had hastened his transition from youth to adulthood, and suffering had made him wise. Now, was the time for him to relax and enjoy the fruits of his toil and hardwork. Then, this deep feeling of life's unfairness, like bile, rises in me. Why did he have to go? There was still so much for him to do. He had to live his life, fulfill his dreams, rejoice and claim his place under the sun.

Hinduism believes in rebirth. We keep coming back from the land of the dead (pitr-lok) to the land of the living (bhu-lok) till we learn the ultimate lesson and merge with the paratman. My deeply spiritual sister offers me solace, 'He was an old soul in a young body. He came here for a purpose and once it was complete, he left. It is poor sods like us who have to complete this whole karmic cycle of life, death, suffering...'

When I heard the song 'Demons' by American pop rock band Imagine Dragons, it resonated with me at so many levels, particularly the following lines:

They say it's what you make
I say it's up to fate
It's woven in my soul
I need to let you go

Your eyes, they shine so bright
I wanna save that light
I can't escape this now
Unless you show me how.

The song expresses so beautifully what I want to tell Jeetu.

Later, I watched the music video of the song in which various characters with personal hardships are depicted, including a grieving daughter, a man with Marfan syndrome, a victim of parental abuse and a military veteran somewhere in the battlefield carrying his buddy (injured or dead?) on his shoulder (the excerpt reproduced above plays in the background). All these characters are there in the concert too, their stories playing out in the background. The video shows how we carry this public persona, behaving normally, bearing these islands of pain and sorrow within: 'Don't get too close, it's dark inside. It's where my demons hide.'

The clip ends with a dedication to Tyler Robinson (1995–2013), a fan of the band who died at the age of seventeen in March 2013 following a battle with cancer.

Sometimes, I felt my relationship with my brother was a sealed sanctum that I did not want anyone to enter. I hid my grief like the polished family silver under my competent professional visage.

I polish my pain
And hide it like family silver
Out of reach, of prying eyes.

Now, speaking with his seniors, colleagues and family almost fifteen years after Jeetu's death, I am finally opening up. And I believe, so is my family. Talking about him, writing about him has freed a caged bird from my soul. Now, I am able to talk about my brother, whose life may have been short in terms of biological years, but who lived many lives in one. On reading my poem 'Monody' people ask—who is this poem about? Still controlling that lump in my throat, now I talk about him. Do I see pity in their eyes or empathy? I cannot figure it.

You were the warrior,
The vanquisher of enemies,
Looking death in the eyes,
Recipient of gallantry awards,
Hazel-eyed and ramrod straight,
Holding untold storms within you.
Three of us were to grow old together.
You left too soon, brother!
Chanting, Charaiveti,
We sisters keep moving, keep moving...

Visiting the NDA and participating in the silver jubilee celebrations in June 2023 was cathartic and painful in equal measure. One was struck by what one had lost, but then there was closure too. Abha was able to visit the house in the campus where she had spent some of the happiest times of her life. Kartikeya was able to see his father's alma mater and hopefully appreciate his father's brief but glorious life. As the mystic poet Rumi says:

Your grief for what you've lost lifts a mirror
Up to where you're bravely working.

Expecting the worst, you look, and instead,
Here's the joyful face you've been wanting to see.

The Indian Army lost a bright officer and we lost a lovely human being. Generally, people do not talk ill of the dead. We revere the dead and talk great things about them. However, speaking to Jeetu's seniors, colleagues, friends, and family members made me realize he was so special and loved by all. He left an impact on all of them and what shone the brightest was his unpretentiousness, equianimity, sincerity, compassion... I can go on and on. The stories that they shared were heartfelt, genuine and imbued with love and affection. There were aspects of his personality that were unique to him and shone bright.

Veins, the long meandering lines of colour, traversing a natural stone give them a distinct character. These veins are the crystallized sheets of minerals deposited in a rock over centuries. Veins made up of intelligence, honesty, integrity, hardwork, and empathy were imprinted on Jeetu's soul and made him the special man that he was. I have tried to use different coloured threads to weave the tapestry of the short life that he spent with us. Here was a young man who lived his life to the fullest, on his own terms, with bravery, dignity and grace.

Salute to you, dear brother, and I hope I have been able to create from our collective sorrow something of a beauty, befitting your life. Wherever you are, Rest in Peace.

List of Trainings Attended by Major Shatrujeet Kotwal, IC-58609H

	Name of the Training and the Institute	Duration of the Training
1	Young Officers (YO-91) Course at the Infantry School, Mhow	08 March 2000 to 31 July 2000
2	Commando Course at the Infantry School, Belgaum	24 October 2000 to 30 November 2000
3	Mountain Warfare Basic Course (MWBC) at the High Altitude Warfare School	30 July 2001 to 15 September 2001
4	Mountain Warfare Advance Course (MWAC) at the High Altitude Warfare School	17 September 2001 to 13 October 2001
5	Winter Warfare Basic Course (WWBC) at the High Altitude Warfare School	06 January 2003 to 08 March 2003
6	Winter Warfare Advance Course (WWAC) at the High Altitude Warfare School	10 March 2003 to 29 March 2003

7	All Arms Officers Combat Engineering Course at the College of Military Engineering	21 March 2005 to 30 April 2005
8	Junior Command Course at the Army War College, Mhow	08 October 2007 to 01 December 2007

Acknowledgements

This book has been possible because of two people—Kiran Kotwal, our mother, and Colonel Dheeraj Kotwal, Jeetu's friend and colleague. Sometimes, I would doubt if I could undertake this whole endeavour to chronicle Jeetu's valiant life, but Mom's unwavering faith in my capabilities kept me going. Dheeraj was the force who opened so many portals for me to reach out to Jeetu's bosses, colleagues and friends. Dheeraj himself was candid in speaking about his decade-long association with Jeetu. It is not easy to be vulnerable. Thanks, Dheeraj.

Thanks Kartikeya for coming into our lives, you were the reason this book was born.

During my conversations with Jeetu's bosses, colleagues and friends, what stood out was their readiness to speak about Jeetu. Surprisingly, their memories of Jeetu were fresh and tinged with love and affection. I do not have enough words to thank his bosses—Colonel Hasabnis and Brigadier Bodhe; his friends and colleagues in the 34 RR—Colonel Bhatt, Colonel Aditya Verma, Lieutenant Colonel Rohan and Lieutenant Colonel Alok Dass (retd.)

for speaking to me and sharing their memories about Jeetu.

It would not have been possible to describe the important events in Jeetu's professional life without Colonel Harjinder Singh, Jeetu's 34 RR CO's inputs. Thank you so much for it. Colonel Singh was unrestrained in sharing his thoughts not only about Jeetu but also about the other two young officers who were Jeetu's colleagues and friends—Major K.P. Vinay and Major Raghu Raman. His words recreated for me the two encounters that led to these three young men winning gallantry awards. Apart from them, Colonel Singh was equally proud of his other officers who were with him in the 34 RR. Similarly, the officers who had worked with Colonel Harjinder remember him as a passionate commanding officer.

While speaking to Jeetu's friends from the Mike Squadron, NDA—Colonel Sharad Kumar Mishra, Colonel Ashwini Mishra, Colonel Biswajit Dash, Colonel Kshitij Balhara and Colonel Arun Nayan, I was transported as if in a time machine to their NDA days. They were like a band of brothers and I got a glimpse of this palpable closeness stretching over years. There are others like Umang Trehan who continue to be in touch. My special thanks to all of you for enclosing me in this circle while recounting your memories. Also, thanks to Colonel Rajeev Bhola, Jeetu's roommate in the IMA for sharing his thoughts on their IMA days.

While speaking about Jeetu, Ami Lal and Ranbir expressed such raw emotions that it brought tears to my eyes.

Ranjeet, Jeetu's school friend was the bridge between his childhood and adulthood. The bond forged between the two was much strong and continues...Thanks for being there Ranjeet.

Vinkoo and Abha were a little sceptical when I broached the idea of the book. However, as my journey took me far and wide, both of them joined with full enthusiasm. They also plumbed their innermost thoughts and memories. It was like diving into this sea of pain that had been dyked but was now required to be crossed. Thanks for being my fellow travellers in writing of this book.

Words cannot describe the role played by Abha's parents, Late Satish Uncle and Veena Aunty in enveloping their youngest daughter and their grandson with so much love and affection. They were the bulwark against the storm. Deepa, Abha's sister and Pawanji, her brother-in-law bestowed much love, warmth and care on Abha and Kartikeya. Their daughters, Bhagyashree and Ojaswani, are the indulgent sisters to Kartikeya. Thank you for being there.

Thanks to Pabitra Chakraborty, my brother-in-law and Vinkoo's husband for his strong and silent support. Anadi Chakraborty, many thanks for refurbishing some of the old photographs. Thanks Arjun for sharing your memories of Jeetu Bhaiya and being part of the family.

Special thanks to Yateendra Joshi for shepherding the book in its initial avatar. It has undergone some changes, but the kernel remains the same.

Thanks to Ravi Singh of Speaking Tiger for having faith in the potential of the book and Jeetu's story. Aruna Ghose through her insightful editing added much value and helped in creating a better version of the book. Dyuti Roy, for the last-mile support before the book is finally published.

Madhu Nag, my husband and Shravani, my daughter bore with me patiently as I ventured out on these multiple journeys. They keep my restless soul anchored.

Last but not the least, my mother-in-law, Lanka Subhadra for her strong support and being my silent cheerleader.

ALSO IN SPEAKING TIGER

MY WORLD WITHOUT JEHAN

Surviving a Brother's Suicide

Liana Mistry

Relocating every few years, not always together, the Chenoys were far from a typical household. For all that, they were a happy family, until they could no longer pretend to be, when, in September 1986, they lost Jehan, the most tender and passionate of the five of them.

Many years later, the kid sister, Liana, tries to fill the gaping hole left by her brother, the loving though tempestuous young man who took his life at twenty-two. She was fourteen at the time. All she has of him now are her memories, fading at the edges, like the handful of old photographs she has of him. She can recall the flamboyant and popular boy at school, but also the one at home, manic depressive, and struggling to come to terms with his sexuality. Her family chose never to speak about the suicide—particularly to their youngest member—instead burying the pain they must have felt. Were they as ridden with guilt as she was, and is, for not being able to prevent Jehan's death? Were they ashamed that he was not like other boys his age, that he was different? She doesn't know. She lives her life grieving privately, imagining they're doing the same.

My World Without Jehan is Liana's cathartic attempt to confront this decades-long silence. Her shattering yet hauntingly beautiful tribute to her beloved brother is ultimately a memoir about forging a path to healing and understanding, and, finally, to letting go.